MIX
Papier aus verantwortungsvollen Quellen
Paper from responsible sources
FSC® C105338

Michaela Meyns

Metal-semiconductor hybrid nanoparticles

Halogen induced shape control,
hybrid synthesis and electrical transport

Anchor Academic
Publishing

Meyns, Michaela: Metal-semiconductor hybrid nanoparticles. Halogen induced shape control, hybrid synthesis and electrical transport, Hamburg, Anchor Academic Publishing 2014

Buch-ISBN: 978-3-95489-293-8
PDF-eBook-ISBN: 978-3-95489-793-3
Druck/Herstellung: Anchor Academic Publishing, Hamburg, 2014

Bibliografische Information der Deutschen Nationalbibliothek:
Die Deutsche Nationalbibliothek verzeichnet diese Publikation in der Deutschen Nationalbibliografie; detaillierte bibliografische Daten sind im Internet über http://dnb.d-nb.de abrufbar.

Bibliographical Information of the German National Library:
The German National Library lists this publication in the German National Bibliography. Detailed bibliographic data can be found at: http://dnb.d-nb.de

All rights reserved. This publication may not be reproduced, stored in a retrieval system or transmitted, in any form or by any means, electronic, mechanical, photocopying, recording or otherwise, without the prior permission of the publishers.

Das Werk einschließlich aller seiner Teile ist urheberrechtlich geschützt. Jede Verwertung außerhalb der Grenzen des Urheberrechtsgesetzes ist ohne Zustimmung des Verlages unzulässig und strafbar. Dies gilt insbesondere für Vervielfältigungen, Übersetzungen, Mikroverfilmungen und die Einspeicherung und Bearbeitung in elektronischen Systemen.

Die Wiedergabe von Gebrauchsnamen, Handelsnamen, Warenbezeichnungen usw. in diesem Werk berechtigt auch ohne besondere Kennzeichnung nicht zu der Annahme, dass solche Namen im Sinne der Warenzeichen- und Markenschutz-Gesetzgebung als frei zu betrachten wären und daher von jedermann benutzt werden dürften.

Die Informationen in diesem Werk wurden mit Sorgfalt erarbeitet. Dennoch können Fehler nicht vollständig ausgeschlossen werden und die Diplomica Verlag GmbH, die Autoren oder Übersetzer übernehmen keine juristische Verantwortung oder irgendeine Haftung für evtl. verbliebene fehlerhafte Angaben und deren Folgen.

Alle Rechte vorbehalten

© Anchor Academic Publishing, Imprint der Diplomica Verlag GmbH
Hermannstal 119k, 22119 Hamburg
http://www.diplomica-verlag.de, Hamburg 2014
Printed in Germany

Metal-semiconductor hybrid nanoparticles: Halogen induced shape control, hybrid synthesis and electrical transport

(Metall-Halbleiter Hybridnanopartikel:
Halogeninduzierte Formkontrolle,
Hybridsynthese und elektrischer Transport)

Dissertation

zur Erlangung des Doktorgrades

im Fachbereich Chemie

der Universität Hamburg

vorgelegt von

Michaela Meyns

aus Reinbek

Hamburg,
April 2014

The experimental work described in this thesis was carried out between April 2010 and December 2013 in the group of PD Dr. Christian Klinke at the Institute of Physical Chemistry of the University of Hamburg.

1. Referee: PD Dr. Christian Klinke
2. Referee: Prof. Dr. Horst Weller

Date of thesis defence: 16 May 2014

Contents

List of Abbreviations   ix

List of Figures   xi

List of Schemes   xiii

List of Tables   xiii

1 Introduction   **1**

2 Halogen induced shape control of CdSe nanoparticles   **5**
    2.1 Properties and synthesis of semiconductor nanoparticles 6
       2.1.1 Properties of semiconductor (CdSe) nanoparticles 6
       2.1.2 Colloidal semiconductor nanoparticle synthesis and shape control . 8
           2.1.2.1 Basics of nanoparticle nucleation 9
           2.1.2.2 Mechanisms of shape control 9
           2.1.2.3 The hot injection synthesis and shape control 13
    2.2 CdSe nanoparticle shape evolution tuned by halogenated additives 15
       2.2.1 Effects of 1,2-dichloroethane on the synthesis of CdSe nanorods . . 15
       2.2.2 Shape and size manipulation with other halogen compounds 20
           2.2.2.1 Variation of the halogen additives 21
           2.2.2.2 Size control by additive injection after the nucleation . . . 24
       2.2.3 Morphology related changes of phosphorous and halogen contents examined by surface and elemental analysis 24
       2.2.4 Ligand-surface interactions and the hexagonal pyramidal shape . . 29
       2.2.5 A closer look at the shape evolution process 34
    2.3 Conclusions 37

3 Metal-semiconductor hybrid nanoparticles   **39**

- 3.1 Properties and synthesis of metal-semiconductor hybrid nanoparticles ... 40
  - 3.1.1 Properties of metal-semiconductor hybrid nanoparticles 40
  - 3.1.2 Synthesis of metal-semiconductor hybrid nanostructures 43
    - 3.1.2.1 Deposition of metallic domains 44
    - 3.1.2.2 Ion exchange 48
- 3.2 Metal-CdSe nanopyramid hybrid structures - deposition and ion exchange . 50
  - 3.2.1 Au-CdSe nanopyramid structures 52
    - 3.2.1.1 Influence of the precursor oxidation state on the hybrid morphology 56
    - 3.2.1.2 Compositional analysis by EDX and XPS 62
    - 3.2.1.3 Reasons for the different deposition behaviour 68
    - 3.2.1.4 Annealing of Au-CdSe shell monolayers 69
  - 3.2.2 Reactions of CdSe nanopyramids with Ag, Pd and Pt 71
    - 3.2.2.1 Reaction with Ag(I) 71
    - 3.2.2.2 Reactions with Pd(II) 73
    - 3.2.2.3 Reactions with Pt(II) 78
- 3.3 Conclusions 85

4 Electrical transport in hybrid nanoparticle films   87
- 4.1 Electrical transport in nanoparticle arrays 88
  - 4.1.1 Transport mechanisms and photoconductivity 90
    - 4.1.1.1 Coulomb blockade and single-electron transistor 90
    - 4.1.1.2 Transport mechanisms in non-crystalline materials 91
    - 4.1.1.3 Photoconductivity 93
  - 4.1.2 Electrical transport in metal-semiconductor hybrid nanoparticles . . 93
    - 4.1.2.1 Macroscopic metal-semiconductor contacts 94
    - 4.1.2.2 Metal-semiconductor nanocontacts 95
- 4.2 Electrical transport in CdSe and Pt-CdSe nanoparticle devices 97
  - 4.2.1 Electrical transport through pyramidal CdSe nanoparticles 98
  - 4.2.2 Electrical transport through Pt-CdSe hybrid nanoparticles 101
    - 4.2.2.1 Electrical transport in Pt-CdSe hybrid nanoparticles with 1.7 nm Pt domains 102
    - 4.2.2.2 Electrical transport in Pt-CdSe hybrid nanoparticles with 3.2 nm Pt domains 103
- 4.3 Conclusions and perspective 111

5 Experimental 113

5.1 Materials and preparation methods 113
5.1.1 Materials 113
5.1.2 Synthesis of pyramidal CdSe nanocrystals 114
5.1.2.1 Standard recipe 114
5.1.2.2 Variation of the halogenated additives 114
5.1.2.3 Injection of 1-chlorooctadecane after CdSe nucleation 115
5.1.2.4 Determination of the relative concentration of protons in aliquots 115
5.1.2.5 Samples for XPS 115
5.1.3 Synthesis of Au-CdSe pyramid hybrid nanoparticles 116
5.1.3.1 Gold(III)-stock solutions 116
5.1.3.2 CdSe dispersions 116
5.1.3.3 Au-CdSe hybdrid nanoparticles with DCE-nanopyramids 117
5.1.3.4 Au-CdSe hybdrid nanoparticles with COD-nanopyramids 118
5.1.4 Reactions of CdSe nanopyramids with Ag, Pd and Pt precursors 120
5.1.4.1 Silver 121
5.1.4.2 Palladium 121
5.1.4.3 Platinum 123
5.1.5 Langmuir-Blodgett monolayer preparation and annealing 124
5.1.5.1 Nanoparticle purification 124
5.1.5.2 Film preparation and annealing 125

5.2 Characterisation 126
5.2.1 Transmission Electron Microscopy (TEM), Energy Dispersive X-ray Spectroscopy (EDX) 126
5.2.2 UV-Visible absorption and fluorescence spectrometry 126
5.2.3 X-ray powder diffraction (XRD) 128
5.2.4 Scanning electron microscopy (SEM) 128
5.2.5 Total Reflection X-ray Fluorescence Spectroscopy (TXRF) 128
5.2.6 Attenuated Total Reflectance Fourier Transformation Infrared Spectroscopy (ATR-FTIR) 129
5.2.7 X-ray Photoelectron Spectroscopy (XPS) 129
5.2.8 Electrical transport 130

6 Summary 131

Contents

7	Zusammenfassung	133
A	Additional data	I
B	Safety	VII
	Bibliography	XXIII
	Acknowledgements	LVII
	Curriculum Vitae	LIX
	Publications and conference contributions	LXI
	Erklärungen	LXIII

List of Abbreviations

BE	Binding energy
CB	Conduction band
COD	1-Chlorooctadecane
Cub.	Cubic
DBE	Dibromoethane
DCE	1,2-Dichloroethane
DDA	Dodecylamine
DDT	Dodecanethiol
DEG	Diethyleneglycol
DFT	Density functional theory
DIE	1,2-Diiodoethane
DTAB	n-Dodecyltrimethylammonium bromide
DTAC	n-Dodecyltrimethylammonium chloride
EDX	Energy dispersive X-ray spectroscopy
FWHM	Full width at half maximum
Hex.	Hexagonal
HDA	Hexadecylamine
HOPG	Highly oriented pyrolitic graphite
ICP-OES	Inductively Coupled Plasma Optical Emission Spectroscopy
MIGS	Metal induced gap state
Monocl.	Monoclinic
NNH	Nearest-neighbour hopping
OAc	Oleic acid
OAm	Oleylamine
Orthorhomb.	Orthorhombic
PMMA	Polamethylmetacrylate
QY	Quantum yield

List of Abbreviations

STEM	Scanning transmission electron microscopy
TBAB	Tetra-n-butylammonium borohydride
TCE	1,1,2-Trichloroethane
TEM	Transmission electron microscopy
Tetr.	Tetragonal
TOP(O)	Tri-n-octylphosphine (oxide)
TXRF	Total reflection X-ray fluorescence spectroscopy
UV-Vis	Ultraviolet-visible
VB	Valence band
VRH	Variable-range hopping
XPS	X-ray Photoelectron Spectroscopy
XRD	Powder X-ray diffractommetry

List of Figures

2.1	Wurtzite crystal structure of CdSe	8
2.2	Types of ligand coordination in nanoparticles	10
2.3	Factors influencing the size and shape of nanoparticles	12
2.4	LaMer plot with separated growth stages	13
2.5	Evolution of CdSe absorption and morphology with the additive DCE	16
2.6	Correlation of nanopyramid size and DCE/Cd ratio	18
2.7	Micrographs of CdSe nanopyramids with oleic instead of phosphonic acid	20
2.8	Temporal evolution of absorption maxima with different chloroalkanes	22
2.9	TEM micrographs of samples prepared with DCE, DBE and DIE	23
2.10	Addition of 1-chlorooctadecane after the nucleation	25
2.11	XPS spectra of nanorods and -pyramids	26
2.12	Temporal changes of the elemental composition determined by TXRF	29
2.13	Rod and hexagonal pyramidal geometry with distinct facets	31
2.14	DFT simulations: PPA on different crystal facets	33
2.15	LaMer-type plot of the CdSe nanopyramid formation reaction	34
2.16	Plot of the pH of aliquots versus time	36
3.1	Band position and work functions of CdSe and different metals	41
3.2	Modes of epitaxial heterodeposition	45
3.3	Mechanisms of oligomer formation	46
3.4	Scheme of reactive sites of CdSe nanopyramids	51
3.5	Micrographs of nanopyramids with cluster sized Au domains	52
3.6	Au domain growth and absorption spectra with increasing Au/CdSe ratio	55
3.7	Photographs of Au precursor solutions	57
3.8	HR-TEM micrographs of CdSe nanopyramids with Au shell and dots	59
3.9	IR-spectra of CdSe nanopyramids before and after treatment with dodecanethiol	60

List of Figures

3.10 X-ray diffraction pattern of Au-CdSe shell nanoparticles 62
3.11 Atomic ratios and diameters of Au-CdSe dot and shell samples 63
3.12 XPS survey spectra of CdSe and Au-CdSe nanoparticles 65
3.13 XPS data of Se 3d and Au 4f regions of Au-CdSe samples with shell and dots. 67
3.14 TEM micrographs of annealed Au-CdSe films 70
3.15 TEM micrograph of nanopyramids after ion exchange with Ag. 72
3.16 Diffraction pattern after ion exchange from Cd to Ag 73
3.17 TEM micrographs of CdSe nanopyramids at different stages of cation exchange with Pd 74
3.18 High resolution micrographs, EDX mapping and electron diffraction of CdSe-Pd_xSe_y nanoparticles 76
3.19 Evolution of absorbance spectra during an ion exchange reaction between Cd and Pd 78
3.20 Oligomeric Pt-CdSe nanoparticles with different Pt domain sizes 79
3.21 Absorbance and emission of Pt-CdSe samples 80
3.22 Electron diffraction and interface regions of Pt-CdSe nanoparticles 81
3.23 EDX map of Pt-CdSe hybrid nanoparticles. 81
3.24 TEM micrographs of *in situ* annealed Pt-CdSe nanoparticles 83
3.25 High resolution micrographs of annealed Pt-CdSe nanoparticles 84

4.1 Scheme of a nanoparticle based source-drain device 88
4.2 Coulomb blockade 90
4.3 Electron transport mechanisms in disordered semiconductors 91
4.4 Band structures and charge transport at metal-semiconductor junctions . . 95
4.5 Interdigitated array electrodes 97
4.6 Micrographs of a CdSe nanopyramid array 99
4.7 Current-voltage curves of a CdSe nanopyramid device 100
4.8 Micrographs of Pt-CdSe nanoparticles and devices 102
4.9 Dark and photocurrent of Pt-CdSe arrays (Pt= 1.7 nm) 103
4.10 Dark and photocurrent of Pt-CdSe arrays (Pt= 3.2 nm) 105
4.11 Fits of current-voltage curves of Pt-CdSe arrays 106
4.12 Temperature dependence of dark currents (Pt= 3.2 nm) 109
4.13 Possible paths for electron transfer in Pt-CdSe arrays 110

5.1 Lengths in hexagonal (di)pyramids 127

A.1	Temporal evolution of absorbance and emission in a reaction with DCE	I
A.2	Morphological evolution without additive	I
A.3	Optical and morphological evolution in reactions with chloroalkanes	II
A.4	Electron beam induced migration in Au-CdSe samples	III
A.5	XPS Se 3d signal of a Au-CdSe shell sample	III
A.6	Beam current dependency of migration in Au-CdSe samples	IV
A.7	Composition of nanoparticles after ion exchange with Ag	IV
A.8	Evolution of absorbance spectra of CdSe nanopyramids incubated with Pd at room temperature	V
A.9	X-ray diffraction pattern of Pt-CdSe nanoparticles	V
A.10	Electrode with CdSe nanopyramid array after annealing.	VI
A.11	Schematic depiction of a 4-point probe measurement	VI
B.1	GHS-pictograms	XXI

List of Schemes

3.1	Au deposition onto CdSe nanoparticles in the presence of amine ligands	53

List of Tables

2.1	Nanoparticle dimensions at different DCE/Cd ratios.	18
2.2	Adsorption energies of ligands on dominant CdSe surfaces	31
3.1	Standard reduction potentials of relevant species	48

List of Tables

3.2 Ionic radii, acid hardness, relevant selenides and their crystal structure(s) of different metal cations 50
3.3 Composition of nanopyramids before and after reaction with Pd(II) 75
3.4 Atomic composition of CdSe and CdSe-Pt samples determined by EDX. . . 79

4.1 Temperature dependencies of selected electrical transport mechanisms . . . 108

5.1 Details of reactions with different halogenated additives 115
5.2 Details of Au deposition reactions 120
5.3 Parameters for incubation of CdSe nanopyramids (DCE) with Pd(II) . . . 121
5.4 Parameters for incubation of CdSe nanopyramids (COD) with Pd(II) . . . 122
5.5 Parameters for incubation of CdSe nanopyramids (DCE) with Pd(II) 2 . . 122

B.1 Safety information for employed substances VII
B.2 All H, EUH and P statements XI

1 Introduction

Based on their outstanding properties, nanoparticles have entered broad areas of research related to (photo)catalysis [1], energy conversion [2], optoelectronics [3], and biomedicine [4, 5]. A major factor determining these properties and the reactivity of nanoparticles is their size and related to it their large surface-to-volume ratio. The small dimensions cause quantum mechanical confinement of electrons and thus altered physical conditions compared to bulk materials [6]. The reduction in size is also advantageous to save precious materials such as catalytic metals. With proceeding climate change and pollution, high hopes rest on developments in solar energy conversion. Nanoparticles offer a variety of solutions for related applications in photovoltaics and photocatalytic conversion of solar into chemical energy.

During the past decades, the preparation of nanostructures consisting of single and multiple components has developed into a tool box for the creation of purpose-designed materials. Especially the colloidal synthesis can be applied to prepare nanoparticles in a large variety of shapes and sizes with comparatively low effort and costs. With the introduction of high temperature preparation methods twenty years ago [7], a way to obtain highly monodisperse nanoparticles was paved. From then on, the control over size and shape of the particles has grown steadily.

Another strong impulse was the selective formation of multicomponent nanostructures that combine materials with different physical properties [8]. In secondary steps other materials can be grown onto prepared nanoparticles with high precision [9, 10]. Metallic nanostructures deposited onto semiconductors, for instance, facilitate the separation of charges that are photogenerated in the semiconductor [11]. This fundamental process is the basis for improved efficiencies in fields such as photocatalytic water splitting [12]. An interesting feature of metal domains on semiconductor nanoparticles is their ability to improve electrical contacts to the latter [13]. This might be utilised to increase the charge transport in semiconductor nanoparticle arrays and affect the photocurrent obtained under illumination.

CHAPTER 1. INTRODUCTION

Creating and characterising such a nanoparticle array requires a high degree of control over all steps involved, from semiconductor synthesis over metal deposition to the final assembly. To reach this control, an understanding of the fundamental processes accompanying each step is necessary. By modulating the shape of the semiconductor material, for example, the number of sites reactive towards the deposition of metals can be varied which affects the whole architecture. Their comparatively large surface results in lowered melting points of nanoparticles and furthermore makes them susceptible for fast dissolution, the adsorption of molecules and reactions with the surrounding medium [14]. To prevent them from coagulation, nanoparticles are coated by a layer of surfactants, also referred to as ligands or stabilisers. Apart from exhibiting a stabilising effect, the adsorption of such ligands may play an important role in the shape evolution of wet-chemically prepared nanostructures. In connection with the preparation of CdSe-carbon nanotube composites 1,2-dichloroethane was observed to induce the formation of hexagonal pyramidally shaped CdSe nanoparticles [15, 16]. A mechanism involving chloride ions was presumed.

The peculiar pyramidal geometry provides a high number of sites prone to metal deposition and is thus an ideal candidate for an envisaged synthesis of hybrid nanoparticles with several defined metal domains in an oligomeric structure. The control of the shape evolution in reactions without carbon nanotubes and a possible adaptation of the method to develop different morphologies is desired. For this reason, a better understanding of the influence of the di-halogen alkane on the shape evolution shall be gained in this work. Another goal is the preparation of hybrid nanoparticles with oligomeric structure for the electrical characterisation in two-dimensional arrays. Finally, the obtained material shall be assembled and pioneering electrical studies shall be conducted to find out how the interplay of the domains influences two-dimensional conductance and the generation of photocurrents.

Following these steps, this thesis is separated into three chapters, each with a theoretical introduction, a results and discussion as well as a conclusions section. In the concluding sections, specific aspects concerning the results of the corresponding chapter will be treated.

In chapter 2, the interactions between ligands and the nanoparticle surface play an important role. The shape control of the semiconductor component by halogenated additives is examined. To better understand why 1,2-dichloroethane induces such a peculiar shape and if the effect may be exploited to tune the size and shape of nanoparticles, experimental and theoretical studies are combined. Systematic variations of halogenated additives as well as elemental and surface analysis are applied. In the theoretical part, calculations based on the density functional theory and parallels to a classical crystal growth model are presented.

In chapter 3, the seeded-growth deposition of four metals, Au, Ag, Pd and Pt, onto hexagonal pyramidal CdSe nanoparticles in organic solution is examined. The reasons for the formation of a presumed shell-like Au structure with undefined composition observed in preliminary work ([17]) are scrutinised. Among the conducted experiments are studies concerning the influence of the oxidation state of the precursor on the morphology of forming hybrids. The oxidation state of the metal in the shell is examined by X-ray photoelectron spectroscopy. The deposition behaviour of the other three metals is tested by a new common synthetic method with oleylamine as ligand and reducing agent for the metal.

Chapter 4 deals with electrical properties of nanoparticles and specifically two-dimensional Pt-CdSe hybrid nanoparticle arrays. Nanoparticles with two sizes of Pt domains are examined. The assemblies are prepared with the Langmuir-Blodgett technique and measured in a probe station under vacuum. Current-voltage curves are recorded in darkness and under illumination conditions with different irradiation sources. Additionally, the temperature is varied down to cryogenic conditions. In comparison to literature data of related systems observations concerning the transport mechanisms are made.

Experimental details of all chapters are summarised in a joined experimental chapter (chapter 5). A general summary with outlook for the work follows in chapter 6, while chapter 7 contains a summary in German. Additional data for chapters 2, 3 and 4 is provided in the appendix.

2 Halogen induced shape control of CdSe nanoparticles

Due to their special size dependent optical and electronic properties semiconductor nanoparticles find application in photovoltaics [18, 19], (photo)catalysis [1, 12], light emitting devices [20] and biological labelling [21, 22], among others [23].

In these contexts, the shape of the nanoparticles may become important due to physical properties depending on the dimensionality of quantum confinement in the nanostructure or simply their packing behaviour [3, 24, 25, 26]. Apart from their morphology, an aspect that is critical for applications is the passivation of the nanoparticle surface by stabilisers.

Owing to the successful implementation of nanoparticles partially capped by or post synthetically treated with halides into solar cells with increased efficiency [27, 28], incorporation of atomic halogen ligands has attracted much attention recently [29, 30, 31, 32, 33]. In addition to enhancing physical properties in nanoparticle arrays, halides show interesting effects on the shaping of nanoparticles. With metal nanoparticles they are deliberately employed to control the growth with influences on the shape formation [34, 35]. For semiconductor nanoparticles, several cases of wurtzite structures with hexagonal bullet, pyramid, pencil or diamond shape were reported which had in common that chloride precursors were present [36, 37, 38]. A few studies showed an increase of morphological uniformity in branched cadmium chalcogenides with wurtzite arms growing on seeds of deviating crystal structures when halides were added [39, 38, 40]. Juárez and co-workers observed how traces of 1,2-dichloroethane employed as solvent for carbon nanotubes added *in situ* to a synthesis of CdSe nanorods induced a shape evolution [15, 16]. Hexagonal pyramidal nanoparticles with wurtzite structure evolved, which then formed composites with the nanotubes. Chloroalkanes were furthermore reported to aid the preparation of sheet-like lead sulfide nanostructures with the chemical structure of the molecules influencing the dimensions of the crystals [41, 42]. These circumstances built a promising foundation for further studies on the shape control of semiconductor nanoparticles by halogen compounds with CdSe as a model system with well-known properties. An extension of feasible morphologies presents an attractive goal with respect to increasing demands on the control of nanoparticle shapes for application in thin film arrays or as seed material with controllable reactive sites for heteronanoparticle formation.

In the following, the most important theoretical aspects of the wet-chemical nanoparticle synthesis and methods of semiconductor nanoparticle shape control will be introduced. They will lay the basis for the discussion of aspects of the shape control of CdSe nanoparticles aided by (organo) halogen compounds examined in this work. For more detailed insights into properties and preparation of nanoparticles, references [23, 43, 44, 45, 46] are suggested. A substantial part of the chapter is based on and reproduced in part with permission from [Meyns, M., Iacono, F., Palencia, C., Geweke, J., Coderch, M. D., Fittschen, U. E. A., Gallego, J. M., Otero, R., Juárez, B. H., Klinke, C. *Chem. Mater.* **2014**, *26*, 1813-1821.] Copyright [2014] American Chemical Society.

2.1 Properties and synthesis of semiconductor nanoparticles

The field of semiconductor nanoparticle synthesis offers a wide and growing variety of synthetic protocols and factors that can be tuned to influence and design the dimensions and shapes of the crystals. Known morphologies range from zero- to two-dimensional in terms of bulk-like dimensions (quantum dots [47], nanotubes and nanowires [48, 49], tetrapods [50], nanosheets [51, 52]). This is appealing in so far as the shape of nanoparticles influences physical and chemical properties through changes in electric fields and crystal facet dependent surface reactivity [14, 53, 54, 55].

2.1.1 Properties of semiconductor (CdSe) nanoparticles

In terms of electrical properties, semiconductors take a middle position between highly conducting metals and insulating materials such as glasses or polymers. The reason for this is that they are able to conduct electricity only when activated by thermal energy or light. This additional energy allows electrons to move from the lower valence to the upper conduction band across the band gap; in metals there is no gap between the bands, whereas insulators are defined as having a band gap bigger than 4 eV. These bands, formed by energetically close lying electron orbitals, are delocalised over the whole crystal and

2.1 Properties and synthesis of semiconductor nanoparticles

excited electron-hole pairs (excitons) are free to move in all directions in bulk materials. In nanoparticles, the number of atoms, each contributing one orbital to the band, is much smaller (between 100 and a few 10000), which causes an increase of the gap with decreasing nanoparticle size and the formation of discrete states at the band edges [6, 24]. At sizes of a few nanometres, the wave functions of electron and hole become confined in the dimensions of the semiconductor nanoparticles, now called quantum dots, and the gap is strongly size dependent (quantum size effect). The number of nanoparticle dimensions ranging in the confinement regime determines the density of available energetic states. This way, the shape and directed growth of a nanostructure may influence its band gap [56]. In spherical nanoparticles all three dimensions are affected and the electronic levels develop into discrete states. This is visible by blue shifts in absorption and emission spectra, in which the first maximum at lower wavelengths in the absorption spectra can be employed to directly relate size and band gap of the nanoparticles [57, 58]. Due to the high surface-to-volume ratio, the configuration of the outer layer of atoms and the ligand sphere protecting the nanoparticles from coagulation make a large impact on the overall properties. For example, the quantum yield of emission and the electrical transport in nanoparticle arrays can be significantly deteriorated by charge trapping sites in form of dangling bonds.

The medium valued direct band gap of II-VI semiconductor CdSe (1.74 eV in bulk [59]) makes it one of the few materials whose nanoparticles can be tuned in size to absorb and emit light throughout the visible spectrum. The electrical properties of CdSe nanoparticles will be addressed in Chapter 4. CdSe crystals occur in the cubic zinc blende or in the hexagonal wurtzite structure. The latter is depicted in Figure 2.1. At room temperature the cubic phase is slightly more stable by 1.4 meV per atom [61]. Polytypism occurs at higher temperatures and the crystal structure depends on the reaction conditions. In the presence of halogens a preferential formation of the wurtzite phase was observed, the reason for which has not been clarified so far [40, 62, 63]. The wurtzite structure is anisotropic, which becomes apparent by differences in the number of bonds per element when counting in opposite directions of the c-axis. In the positive direction Cd has one bond pointing upwards whereas Se has three. For this reason, hexagonal CdSe crystals are usually terminated by a positively charged Cd-rich (0001) and a negatively charged Se-rich (000$\bar{1}$) facet to minimise the number of dangling bonds [64, 65, 66]. This anisotropic charge distribution creates a dipole moment in the nanoparticles [67, 68].

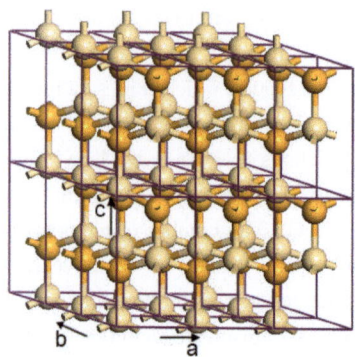

Figure 2.1: Hexagonal wurtzite crystal structure of CdSe composed from unit cells with the lattice constants a, b: 4.30 Å, c: 7.01 Å [60]; Cd atoms are beige, Se atoms are orange. The upper (0001) layer consists of Cd-atoms with one dangling bond, the lower (000$\bar{1}$) facet is usually terminated by Se with one dangling bond. Created with *Materials studio*.

CdSe nanoparticles alone or as cores coated with inorganic or organic shell materials, are applied in solar cells [18], light emitting diodes [20], and biological imaging [4]. Furthermore, they have acted as model systems for the exploration of properties in a variety of contexts. Among these are optical and electrical studies as well as investigations of shape control during synthesis [69, 70, 71].

2.1.2 Colloidal semiconductor nanoparticle synthesis and shape control

For further processing and efficient applications nanoparticles should be homogeneous in their size and shape. The method of choice to obtain nanoparticles with the highest precision in terms of size and shape is the wet-chemical bottom-up approach, where crystals are prepared from molecular precursors of the components. Semiconductor nanoparticles with small size distributions of below 10% are mainly obtained by hot-injection syntheses with separated nucleation and growth stages as pioneered by Murray, Norris and Bawendi in 1993 [7]. Before going into detail, the basics of nanoparticle formation and different models of shape control will be presented briefly.

2.1 Properties and synthesis of semiconductor nanoparticles

2.1.2.1 Basics of nanoparticle nucleation

For the formation of nanoparticles starting materials are mixed and react to a solute form of the crystal[1], not yet considered as a solid phase, which evolves into nuclei once a critical supersaturation is reached. The activation energy necessary for nucleation is provided when the change of the overall free energy

$$\Delta G = -\frac{4}{V_{m,NP}}\pi r^3 k_B T \ln(S) + 4\pi r^2 \gamma \tag{2.1}$$

reaches a maximum. In Equation (2.1) $V_{m,NP}$ is the molecular volume of the material in the crystal, r is the radius of the spherical nuclei, k_B is the Boltzmann constant, S is the saturation ratio and γ is the surface free energy per unit surface area [14]. The maximum is reached for a critical radius r^* when the saturation ratio $S = \dfrac{c_{solute}}{k_{sp}}$ with the solute concentration c_{solute} and the solubility product of the crystal material k_{sp} is larger than one [72]. The critical radius is obtained as

$$r^* = \frac{2V_{m,NP}\gamma}{3k_B T \ln(S)} \tag{2.2}$$

by derivation of Equation 2.1. During nucleation, the supersaturation depletes until no further nuclei form. These nuclei continue to grow towards nanoparticles whose size and shape can be controlled by growth parameters such as concentration, temperature and the choice of ligands. Finally, the solute concentration will approach the level of the material's solubility and the critical radius will shift to larger values so that smaller particles dissolve in favour of further growth of bigger ones, a process known as Ostwald ripening [73].

2.1.2.2 Mechanisms of shape control

Non-spherical shapes form whenever the facets of a crystal exhibit different growth rates; fast growing facets are eliminated and slowly growing facets eventually determine the shape or *Tracht*, the entirety of the facets that form the crystal surface. Growth rates can be governed by thermodynamic or kinetic influences, depending on the growth conditions. Apart from temperature and concentration, a major role in size and shape control is played by

---

[1] In literature the terms *solute* and *monomer* are employed to describe a small unit consisting of at least one of each crystal components. Solute is the term found in works on classical crystal growth and is less limited, since the solute form must not necessarily be a single formula unit.

organic surfactants, which are employed to stabilize the solid nanoparticle phase against dissolution and coagulation [74, 75, 76]. The most commonly employed types of ligands to stabilise semiconductor nanoparticles are amphiphilic organic molecules counted to the groups of neutral L-type or negatively charged X-type ligands. The first group may coordinate both cationic and anionic crystal components, while the second one selectively balances positively charged surface atoms to reach overall charge neutrality. If both types are present, the interaction with X-type ligands will dominate due to stronger interactions with the surface [29, 31, 77, 78]. L-type ligands provide additional stability but are more easily washed away during purification. A breakthrough observation regarding selective X-type adsorption is the case of phosphonic acid impurities in a CdSe synthesis with otherwise neutral tri-n-octylphosphane (TOP) ligands in tri-n-octylphosphane oxide (TOPO), which coordinated the nanoparticles in form of phosphonates [79]. As mentioned earlier, halide ions acting as atomic X-type ligands gain increasing popularity.

The terms *surfactant*, *stabiliser* and *ligand* are often used interchangeably in the literature. Since surfactants and stabilising molecules usually are amphiphilic compounds with alkyl chains, the term ligand will be applied here within the meaning of a potentially surface bound constituent including short molecules and halides.

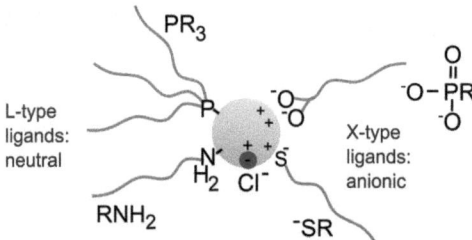

Figure 2.2: Most of the ligands employed in semiconductor nanoparticle stabilisation belong to the groups of neutral L-type ligands, which bind to cations and anions of the crystal (a dative bond to the metal cation is shown here), and X-type ligands that selectively bind to the cationic component. The named examples are tri-n-alkylphosphanes (R_3P), primary amines (RNH_2), thiolates (RS^-) and phosphonates (($RPO_3)^{2-}$). Halides can act as atomic X-type ligands. In metallic nanoparticles with negative surface charges cationic ligands are another important group of surfactants.

Kumar and Nann summarised theories explaining the shape evolution of nanoparticles and named five important ones: (i) thermodynamic theory, (ii) selective-adsorption model, (iii) oriented attachment model, (iv) molecular template theory and (v) effective-monomer model [76].

(i) Thermodynamic theory

According to the thermodynamic theory, which is based on the works of Gibbs, Curie and Wulff, the shape of a crystal in equilibrium with its environment is determined by crystallographic facets whose combined surface free energy minimises the total free energy of the system [80]. In a supersaturated solution equilibrium shapes with low index facets would form because the growth of higher index facets is accelerated. In classical crystal growth theory, thermodynamic control is one approach to understand the *Habitus*[2] of polyhedral equilibrium forms of crystals [81]. An experimentally verified model to predict such forms was developed by Wolff and co-workers [66]. It focuses on the contribution of dangling bonds to the surface free energy of a facet and thus includes their coordination and their ionicity in form of the ratio between cationic and anionic bond energies into the calculation. The traditional thermodynamic theory often does not account for shape evolutions in off-equilibrium systems applied in nanoparticle synthesis. Furthermore, the Wolff model has not been related with modern nanoparticle synthesis and shapes so far, to the best of the author's knowledge.

(ii) Selective-adsorption model

The selective-adsorption model was among the first to explain anisotropic growth of II-VI semiconductor nanoparticles such as CdSe. It is based on the assumption that ligands preferably adsorb to certain crystallographic facets. With a mixture of ligands the different affinities thus influence the relative growth rates of evolving facets, which leads to anisotropic shapes [75, 82].

(iii) Oriented attachment model

A reduction of surface energy can be evoked by attachment of already formed nanoparticles via crystallographically similar facets. This process can be explained by thermodynamic driving forces or dipole-dipole interactions [83, 84]. Such oriented attachment can lead to

---

[2] The term *Habitus* should not be mixed up with *Tracht*. A *Tracht* is the set of crystal facets terminating the surface of a crystal, the *Habitus* is determined by the relative size of these facets.

one-dimensional structures including rods [85] and wires [86] and is thought to play a role in the formation of two-dimensional nanosheets [41].

(iv) Molecular template theory
Organic molecules that are able to coordinate crystal components are employed to form templates for nanoparticle growth. A variety of morphologies from cubes to wires and branched structures of PbSe was thus prepared by employing alkyldiamines as a "solvent coordinating molecular template" [87]. Especially in Cd-chalcogenides alkylamines are capable of inducing the formation of lamellar templates for the synthesis of two-dimensional nanoribbons and quantum belts [52, 88]. The formation and re-crystallisation of CdSe clusters along the template was observed in an intermediate stage and the mechanism was proposed to be generally relevant for the growth of two-dimensional morphologies.

(v) Effective-monomer model
A high monomer concentration after the nucleation characterises reactions that can be described by the effective-monomer model. Here, according to Kumar and Nann, it is not the interaction between ligands and crystal facets but their ability to provide a large monomer/solute concentration after the nucleation stage which controls the shape evolution. A high precursor complex stability is necessary to promote such kinetically driven growth, which causes crystal facets with the highest potential to grow with the fastest rate and thus leads to anisotropic growth. Prominent examples for this kind of control are works by X. Peng and co-workers [71, 89, 90, 91, 92].

Due to the complexity of nanoparticle synthesis it is not always straightforward to ascribe a shape evolution to a certain model. A reaction may undergo different stages with changing shapes, depending on the conditions. The critical, interrelated factors that generally determine nanoparticle growth and shape are summarised in Figure 2.3.

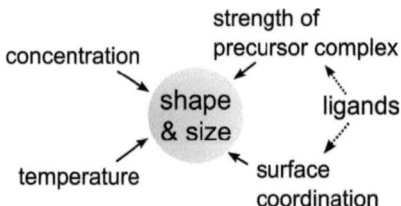

Figure 2.3: Factors influencing the size and shape of nanoparticles during synthesis.

2.1.2.3 The hot injection synthesis and shape control

In order to obtain monodisperse semiconductor nanoparticles, it is important to prevent secondary nucleation events and Ostwald ripening processes. A way to achieve this is by a controlled thermal decomposition of organometallic precursor compounds and nanoparticle growth at high temperatures followed by quick quenching of the reaction. The injection of a room temperature solution containing one of the precursors into a hot solution of the second component results in a short nucleation phase by inducing a quick rise and drop of the supersaturation. A standard model to visualise the correlation of different reaction stages and solute concentration or supersaturation was first proposed by LaMer and Dinegar [93]. The model includes three stages beginning with a build up of supersaturation (I) followed by nucleation (II) and growth. Figure 2.4 shows an extendend version of the traditional plot with the supersaturation instead of the concentration plotted against time and a distinction between different stages of growth (III/IV) [72]. The rate of solute formation from precursor molecules and the solubility of the solute control the number and size of nuclei as well as the degree of supersaturation after the nucleation [94, 95]. A faster solute formation and lower solubility result in a higher supersaturation and, according to Equation 2.2, in smaller nanoparticles. If the solute concentration remains high after the nucleation, diffusion controlled growth makes smaller nanoparticles grow faster than larger ones which results in narrowing of the size distribution (focusing, stage III) [75]. This is also referred to as *kinetic shape control*. If the supersaturation is low, broad size distribu-

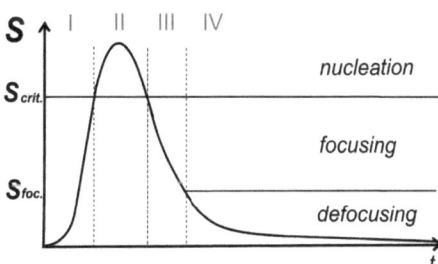

Figure 2.4: LaMer-type plot of the supersaturation S against time t with different stages of the reaction. After the injection a supersaturation develops (stage I) which is partially relieved by nucleation above the critical supersaturation $S_{crit.}$ (stage II). As the supersaturation depletes towards the saturation level, the formed nuclei grow to a more homogeneous size (focusing, stage III) until the next threshold ($S_{foc.}$) below which the system moves towards an equilibrium state and ripening processes occur (defocusing, stage IV) [72].

tions develop early due to Ostwald ripening. Otherwise, ripening sets in during the second stage of growth (defocusing, stage IV).

The evolution of nanoparticle shapes in this type of synthesis is also related to the different reaction stages. With a high supersaturation in stage III the condition for kinetically controlled growth after the effective-monomer model is fulfilled and monodisperse dot- and rod-shaped or tetrapodal nanoparticles form [72]. A high solute/monomer concentration and rod growth of CdSe, for example, were obtained by employing phosphonic acids [89, 90, 92], which form stable complexes with Cd in the precursor and on the nanoparticle surface [77, 96, 97]. Here, the effective-monomer model and selective adhesion of ligands overlap, since the Se-terminated $(000\bar{1})$ facet in CdSe is the one least capped by ligands in the corresponding synthetic conditions, so that nanorods growing preferentially in the $[000\bar{1}]$ direction along the c-axis are formed [98]. If the supersaturation is not maintained at a high level by additional precursor injection or suitable precursor conversion kinetics [75, 79, 99], the depletion of precursor and solute will lead to ripening of the particles which results in the formation of thermodynamically more stable shapes. Peng *et al.* reported a supersaturation dependent transition from anisotropic (1D) to three-dimensional growth followed by intraparticle ripening [90]. Nanorods re-arranged to spherical shapes in a limited diffusion sphere around them without mass exchange with the bulk solution. Variations of the monomer concentration in the growth solution may lead to different morphologies from dots to tetrapods under otherwise identical reaction conditions [71]. The understanding of this type of reaction is broad but due to the complexity of interactions between different components and the sometimes large influence of impurities, for example inducing branching [100], new aspects continue to be revealed [101, 102, 103, 104].

In the following, the influences of varied halogenated additives on the synthesis of CdSe nanorods and possibilities to exploit them will add to this knowledge and reveal a new degree of freedom in nanoparticle shape control.

2.2 CdSe nanoparticle shape evolution tuned by halogenated additives

To find out more about the role of 1,2-dichloroethane in the formation of CdSe nanopyramids and the effect of employing other halogen compounds, synthetic studies with varied additives were combined with surface and compositional analysis and complemented with density functional theory calculations. Central questions to be answered were: 1) why does a hexagonal pyramidal morphology form and 2) is there a way to influence the shape deliberately? After the examinations of different aspects the shape evolution will be discussed based on the findings.

2.2.1 Effects of 1,2-dichloroethane on the synthesis of CdSe nanorods

In the high temperature wet-chemical preparation of CdSe-carbon allotrope composites, residues of 1,2-dichloroethane (DCE) were shown to evoke a morphological transformation of *in situ* prepared CdSe nanorods to hexagonal dipyramidal particles (CdSe pyramids) with wurtzite structure during the course of the reaction [15, 16]. The halogenated solvent had entered the reaction mixture as dispersant for the injected carbon allotropes and was in large part removed *in vacuo*. A similar shape evolution could be obtained by adding aqueous hydrochloric acid to the nanorod-CNT reaction. This method was modified to a protocol without carbon allotropes, which sped up the morphological evolution [17, 105]. A hindrance by carbon allotropes may be explained by disturbed diffusion flux and/or adsorptive interactions of necessary components with carbon.

In a typical CdSe nanopyramid synthesis, DCE was injected to a complex of cadmium with *n*-octadecylphosphonic acid (ODPA) in tri-*n*-octylphosphane oxide (TOPO) at 80 °C before the injection of selenium in trioctylphosphane (TOPSe, 1 M, injected at 265 °C) initiated the formation of nanoparticles which were left to grow at 255 °C for 4 hours (molar ratios: Cd/Se/ODPA/DCE = 1:2:2:0.7). In Figure 2.5, the evolution of the first absorption maximum of samples from a synthesis with DCE is plotted together with data from a control experiment without additive (the full absorbance spectra can be found in Appendix A). The fast shift of the absorption during the first 10 minutes can be correlated with a kinetically controlled stage of rod growth. Afterwards, the shift slew down and the direction of growth changed. In the provided transmission electron microscopy (TEM) micrographs the nanoparticles have grown perpendicularly to the *c*-axis with aspect ratios approaching one and gradually eliminated the flat $\{10\bar{1}0\}$ side facets to give

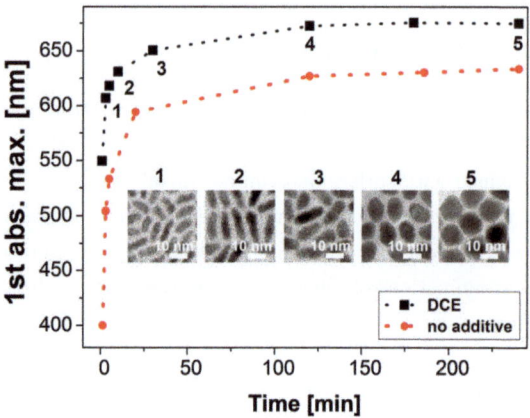

Figure 2.5: Temporal dependence of the first absorption maximum in reactions with and without 1,2-dichloroethane (DCE) and the morphological evolution of nanopyramids with DCE in TEM micrographs. The aspect ratio of the nanoparticles develops from around three after 10 minutes to approximately one after 240 minutes.

hemimorphic hexagonal dipyramids with a (0001) base and dominant $\{10\bar{1}\bar{1}\}$ facets [106]. From this, it can be deduced that the change of growth rates of the facets is caused by thermodynamically controlled ripening processes occurring with the depletion of the supersaturation at proceeding reaction times [79]. Recent studies on nucleation and growth of nanoparticles suggest that the separation of growth stages is not as clear as described earlier due to fluctuations in concentrations [107]. Nevertheless, the current reaction seems to be well described by the four-stage model with solute formation (induction period, I), nucleation (II), kinetically controlled rod growth (III) and thermodynamically controlled ripening to pyramidal nanoparticles (IV). Together with the absorption features the emission shifted towards longer wavelengths with time, accompanied by a broadening of both due to a reduction of confinement with larger nanoparticle sizes [105] (see spectra in Appendix A, compare to reaction with carbon nanotubes [106]). A decrease of the emission intensity is noted with proceeding reaction time. Final quantum yields measured against Rhodamine 6G were below 1% for purified samples. Such low quantum yields often occur with comparatively large CdSe nanoparticles and indicate that the final pyramidal shape exhibits a high number of surface defects (trap states) [108, 109].

2.2 CdSe nanoparticle shape evolution tuned by halogenated additives

In a synthesis without DCE, the absorption wavelength and with it the size of the nanoparticles is shifted to smaller values. The shape develops in the way described by Peng and co-workers, with rods growing one- and later three-dimensionally. Ripening to pyramidal shapes was not observed in the control reaction even after 70 hours when Ostwald ripening had broadened the size and shape distribution (see Appendix A).

By changing the amount of DCE in the reaction, a comparison of DCE/Cd ratios from 0.3 to 1.0 revealed two tendencies. With increasing amount of the additive, (1) the size of rods, both diameter and c-axis, after 10 minutes became larger and (2) the size of the rods is related to the size of the forming nanopyramids, in which the pyramidal morphology was more sharply facetted with more DCE. Micrographs after 10 and 240 minutes with histograms of the shorter axis (diameter) after 240 minutes are depicted in Figure 2.6 together with the temporal evolution of the first absorption maxima during the first stage of growth. The corresponding nanoparticle dimensions are listed in comparison with others in Table 2.1. In UV-Vis spectra, the stage of rod growth was accompanied by a rapid red shift of the absorption maxima in the beginning. In contrast to DCE/Cd ratios of 0.3 and 0.7, the solution remained colourless until 4 minutes after the injection of Se when a ratio of 1.0 was employed. However, a fast red-shift occurred afterwards with only minor shifts after 10 minutes. The slope between absorption maxima is also smaller with DCE/Cd 0.7 than with 0.3, which can be interpreted as a faster transition to the ripening stage with higher concentrations of the chlorinated additive.

The delay observed with the ratio of 1.0 may be caused by an inhibition of particle formation and fast, less defined growth once the nuclei were formed. Instead of smooth $\{10\bar{1}0\}$ side facets the rods preferentially exhibited zigzag shaped ones, indicating a strong influence of DCE on the surface of growing rods at this DCE/Cd ratio. On comparing the micrographs after 10 and 240 minutes it becomes apparent that the irregular shape, size and size distribution of the initial rods influences the quality of the later formed pyramids. With a higher irregularity of the rods, the number of stacking faults (visible as stripes perpendicular to the c-axis of the nanoparticles), and the inhomogeneities in shape and size of pyramids were more pronounced. With a low amount of DCE, on the contrary, tendencies towards a pyramidal morphology had developed only in a few nanoparticles, irrespective of size and thus seemingly random. The ratio of 0.7 seems to be sufficient to promote faceting of all nanoparticles whereas it is low enough to prevent the formation of irregular rods and defect loaded pyramids.

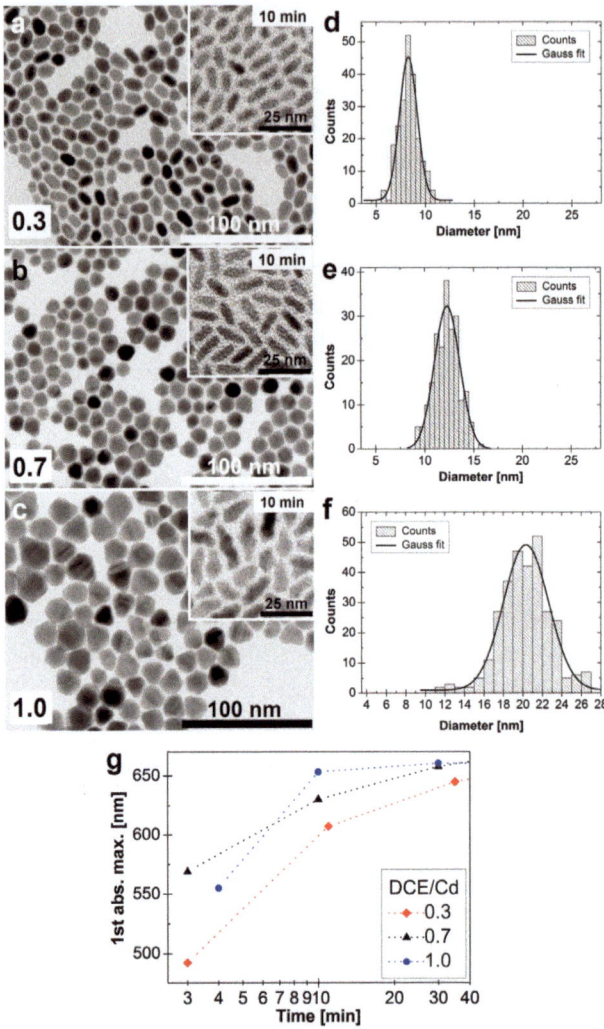

Figure 2.6: (a - c) TEM micrographs after 240 (large) and 10 minutes (inset) and (d - f) histograms of the diameter of samples after 240 minutes, prepared with different DCE/Cd ratios. (g) Evolution of the first absorption maxima during early periods of the reaction.

Table 2.1: Nanoparticle dimensions at different DCE/Cd ratios after 10 and 240 minutes and with CdCl$_2$ instead of CdO as precursor, measured from micrographs (TEM) or powder X-ray diffractometry (XRD).

DCE/Cd	Time	Method	c-axis [nm]	Diameter [nm]
0.3	10 min	TEM	10.4 ± 1.1	4.4 ± 0.4
	240 min	TEM	12.8 ± 2.0	8.3 ± 1.0
0.7	10 min	TEM	13.0 ± 1.3	4.7 ± 0.5
	240 min	TEM	13.1 ± 1.4	12.2 ± 1.3
1.0	10 min	TEM	16.8 ± 2.6	7.4 ± 0.9
	240 min	TEM	20.8 ± 2.3	20.9 ± 2.4
1.3	240 min	XRD	45.8 ± 0.1	52.2 ± 1.6
6.5			no nucleation	
CdCl$_2$	240 min	XRD	34.9 ± 0.5	36.7 ± 0.1

To round off the picture, higher DCE/Cd ratios were tested and the Cd salt (CdO) was exchanged for CdCl$_2$ to be Cd- and Cl-source in one [110]. A ratio of 1.3 evoked a result similar to CdCl$_2$ because in both cases wurtzite CdSe formed but instead of a colloidal solution bulk like precipitate was obtained. Table 2.1 provides an overview of the nanoparticle dimensions, with values well above 30 nm for components of the precipitates. A DCE/Cd ratio of 6.5 was enough to prevent nucleation. These findings comply with earlier attempts to use CdCl$_2$ or salts of other strong acids as precursors, where the salts were highly soluble in the reaction mixture but nanoparticle nucleation could not be effected [90, 111]. Accordingly, the solubility of the reaction components seems to be increased by DCE or chloride and to thus play a significant role in determining the size of the nanoparticles. Earlier nuclear magnetic resonance experiments showed that the Cd-ODPA complex is not observable under high excess of DCE [16]. Thinking of the LaMer plot, an increase of precursor or solute solubility shifts the saturation and critical value for nucleation upwards for a given metal or chalcogenide compound. At the same time, the maximum supersaturation only moves slightly so that the nucleation window is reduced. A higher concentration of ligands, for instance, leads to the formation of larger nanoparticles and, due to the inverse proportionality of the number and size of the produced nanoparticles at constant initial precursor concentration, to a smaller number of nuclei [94, 107, 112, 113, 114, 115, 116]. This line of argumentation complies with other authors who attributed increased Cd-chalcogenide dimensions obtained with higher amounts of halides in solution to a formation of well soluble Cd-halide compounds or mixed complexes with halides and the original ligands [38, 117].

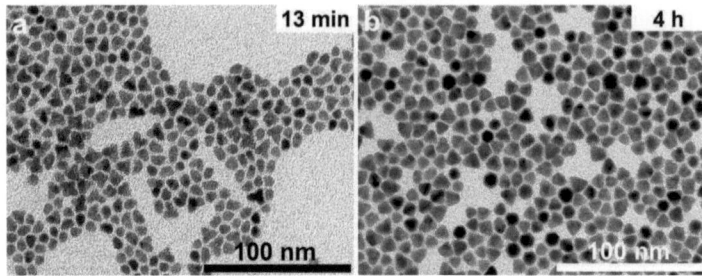

Figure 2.7: With oleic acid instead of octadecyl phosphonic acid pyramidal shapes evolved under the influence of DCE but a high number of tetrapods is visible in the micrograph after 13 minutes, which results in a mixture of tetragonal and hexagonal morphologies after 4 hours.

The influence of DCE on the morphology is not limited to the case with octadecylphosphonic acid as ligand, since pyramidal shapes with wurtzite structure were also observed when oleic acid was employed to complex Cd[3]. Cd-fatty acid complexes exhibit a higher lability and faster nucleation kinetics than the corresponding phosphonate compounds [118]. In the present reaction, this is reflected in the circumstance that the solution turns brown immediately after the injection of Se, omitting the usual induction time and change from colourless via yellow and orange colours. In addition to hexagonal structures tetragons are visible in the micrographs in Figure 2.7, which can be traced back to tetrapodal nanoparticles formed in the first stage of growth. Tetrapods grow preferably from zinc blende nuclei [119], indicating that DCE has a general effect on the expression of surfaces diagonal to the c-axis and the nanoparticle size but ODPA is still the component responsible for anisotropy and the formation of defined rods.

2.2.2 Shape and size manipulation with other halogen compounds

The circumstance that the amount of DCE influences the size of nanoparticles similarly to halides and that the pyramid formation is possible with added hydrochloric acid strongly suggests that the active species is related to chloride (Cl^-) and that different additives may be utilised. It could indeed be shown that the progression of the reaction and thus the shape evolution can be regulated by the choice of additive. For this, chemical analoga in form of different chloroalkanes and dihaloalkanes with bromine and iodine were compared.

---

[3] The ligand to Cd ratio had to be increased to 2.7 in this reaction to fully dissolve CdO.

2.2.2.1 Variation of the halogen additives

A variety of chloro components can be applied to induce shape transformations during the synthesis of CdSe nanorods. Apart from halo alkanes, ammonium salts such as ammonium chloride and n-dodecyltrimethylammonium chloride (DTAC) are capable of evoking the formation of nanopyramids and -bullets[4], thus supporting the hypothesis of chloride as active species [110]. There are two major mechanisms thinkable for the release of chlorine from 1,2-dichloroethane. The first one is an elimination of Cl$^-$ or hydrochloric acid. This can be thermally activated and is catalysed by transition metal Lewis components so that process temperatures that are usually above 340 °C may be lowered to regimes of the present reaction temperature [121, 122]. Acting as Lewis acids, Cd ions in precursor complexes in solution and on the surface of already formed nanoparticles might be able to ease the Cl$^-$ elimination reaction by "pulling" on Cl atoms in the molecule, while traces of water as well as other reaction components may act as proton acceptor. Alternatively, Lim *et al.* proposed a release of halide through the formation of $(R_4P^+(halide)^-)$-type compounds by a reaction between the haloalkane and tri-n-octylphosphane. Recently published results of Cristina Palencia support this latter idea for the nanopyramid reaction [123].

A relation between the molecular structure of the additive and its impact on the nanoparticle shape could be demonstrated by a variety of haloalkanes. Reactions were carried out by substituting DCE in equimolar amounts (halo alkane/Cd: 0.7) with 1,1,2-trichloroethane (TCE), 1,2-dichlorobutane (1,2-DCB), 2,3-dichlorobutane (2,3-DCB) and 1-chlorooctadecane (COD). The additives were injected to the Cd-ODPA complex 10 °C below their boiling point or in the case of 1-chlorooctadecane (boiling point: 348 °C) directly at 265 °C. Again, aliquots were examined by UV-Vis spectroscopy and TEM as shown in Figure 2.8 and in Appendix A, where micrographs and absorption data for the full reaction time are provided. While all reactions yielded pyramidal nanoparticles, differences in the temporal evolution of absorption and morphology as well as the homogeneity of the samples were recorded. The influence on the nanoparticles and thus the reactivity of the additives seems to follow a structure related tendency that would be expected for a substitution reaction. In the beginning of the growth process (after 3 minutes), the wavelengths of the absorption maxima are higher and the nanorods bigger (see micrographs in Appendix A) with additives containing Cl atoms that are less stabilised by neighbouring alkane groups (+I effect and steric stabilization). Similar to the absorption data shown

---

[4] The bullet morphology obtained with DTAC might be caused by amine ligands, which can be formed as side products through its thermal decomposition [120].

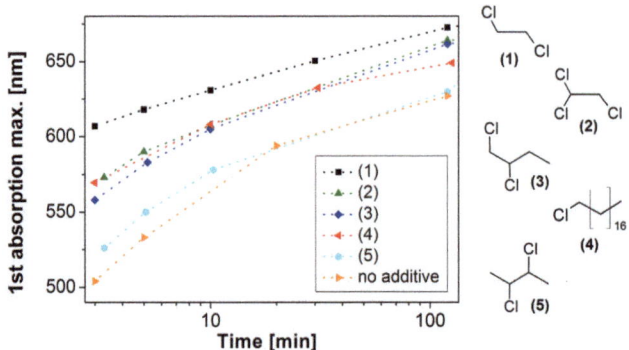

Figure 2.8: Temporal evolution of the first absorption maxima in reactions with chloroalkanes (1 - 5) and without additive. The employed chloroalkanes are (1) 1,2-dichloroethane (DCE), (2) 1,1,2-trichloroethane (TCE), (3) 1,2-dichlorobutane (1,2-DCB), (4) 1-chlorooctadecane (COD) and (5) 2,3-dichlorobutane (2,3-DCB). Longer absorption wavelengths in the beginning (after 3 minutes) indicate a formation of bigger nanoparticles with additives containing sterically less stabilised chlorine atoms. Modified with permission from [124]. Copyright 2014 American Chemical Society.

in Figure 2.5 and Figure 2.6 g, this can be explained by an increased precursor or solute solubility through released Cl^-. The latter causes growth to supersede nucleation at an earlier point and increases nanoparticle sizes [94]. In terms of the shape evolution during ripening, nanopyramids with absorption maxima ranging from 661 nm to 675 nm after 240 minutes were eventually obtained with all additives. With DCE they were most faceted and homogeneous, whereas more rounded shapes or flattened pyramids appeared with other additives such as 2,3-DCB and COD (micrographs in Appendix A). Changes between the measured absorption maxima are less pronounced with the large DCE-nanoparticles than with those prepared with 2,3-DCB, especially in the beginning of the reaction. Between 10 and 240 minutes the shift is 44 nm with DCE and 87 nm with 2,3-DCB. The shifts with the other additives lay between these values. A smaller shift and the more faceted morphology of nanoparticles that were large in the beginning indicate a higher tendency to ripening processes in the monitored period compared to nanoparticles that were smaller in the beginning. In combination with the results obtained with varied amounts of DCE, it can be concluded that a higher accessibility or availability of Cl^- results in larger nanorods and more homogeneously sized and faceted nanopyramids.

2.2 CdSe nanoparticle shape evolution tuned by halogenated additives

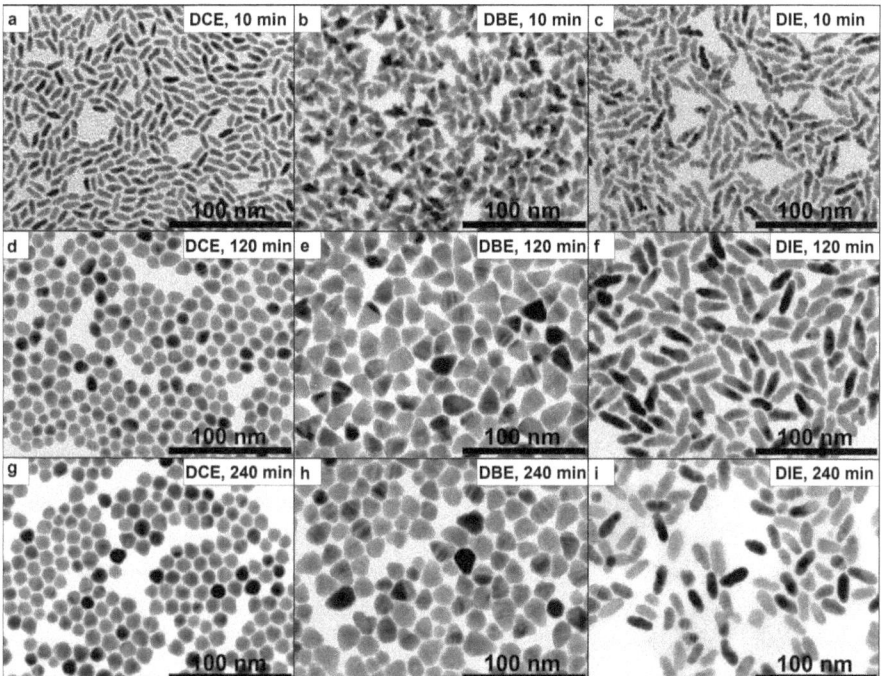

Figure 2.9: TEM micrographs of samples prepared with DCE (a, d, g), DBE (b, e, h) and DIE (c, f, i) at different stages of the reaction. With heavier halogens a tendency to the formation of diagonal facets and increased lengths is visible, while the evolution to pyramidal shapes is less defined or hardly recognisable and the particles exhibit a high number of stacking faults. Reprinted with permission from [124]. Copyright 2014 American Chemical Society.

The hypothesis of a substitutional release of halide which influences the extend to which the shape evolution is affected was further corroborated by experiments with chemical analogues of DCE containing other halogens, in particular 1,2-dibromoethane (DBE) and 1,2-diiodoethane (DIE). The additives were introduced in equimolar amounts (DXE/Cd 0.7). In the kinetically controlled growth regime, anisotropic nanoparticles terminated with zigzag shaped side facets similar to the ones obtained with a DCE/Cd ratio of 1.0 evolved. With both additives the c-axes were even longer (>20 nm) than those observed in the case of DCE. After 10 minutes (TEM: Figure 2.9 a - c), the size increases from nanoparticles with DCE to DBE to DIE, which suggests a stronger influence of the iodoalkane followed by the

CHAPTER 2. HALOGEN INDUCED SHAPE CONTROL OF CDSE NANOPARTICLES

bromoalkane during nucleation and early stages of the reaction compared to chloroalkanes. Thus, the effects on the nanoparticle morphology in this phase of growth must be related to the reactivity of the haloalkanes, determined by both the molecular structure and the nucleophilicity or leaving group ability of their halogen atoms.

When approaching the ripening stage, a trend toward pyramid formation is visible with nanoparticles prepared under addition of DBE, whereas the surfaces of nanoparticles prepared with DIE are smooth and only a slight tendency of growth perpendicular to the c-axis is apparent. This means that the heavier halogens boost the growth process during rod formation but are less effective in the re-shaping process during ripening. The reason for the latter will be discussed in more detail in the next sections.

2.2.2.2 Size control by additive injection after the nucleation

In order to reduce the size of the nanopyramids, the relation between rod and pyramid size can be exploited. If the additive is injected after the nucleation and an initial rod growth, its effect on their dimensions is diminished. Furthermore, the temperature at the moment of Se injection may be increased, which reduces the critical radius of the nuclei and leads to even smaller rods. In presence of additives this would be counter-productive because of an increased release of halide.

An ideal additive for this method is 1-chlorooctadecane with its boiling point well above the growth temperature of 255 °C. At an additive/Cd ratio of 0.7, already formed rods with a c-axis of 5.6 ± 0.7 nm gradually evolved into comparatively small and smoothly shaped hexagonal pyramids with a c-axis length of 7.8 ± 0.7 nm and a small size distribution, even after 24 hours. The corresponding micrographs and absorbance spectra are shown in Figure 2.10. A difference to the nanopyramids prepared with DCE is that the frustum of the pyramids is smaller or non-existent (see model in (c)), which reduces the number of exposed crystallographic vertices and thus reactive sites.

2.2.3 Morphology related changes of phosphorous and halogen contents examined by surface and elemental analysis

Surface analysis of the nanoparticles was carried out by X-ray Photoelectron Spectroscopy (XPS) in collaboration with Fabiola Iacono, Roberto Otero and José M. Gallego from the Universidad Autónoma de Madrid and IMDEA Nanoscience Madrid. XPS spectra were obtained at the BESSYII synchrotron storage ring, Helmholtz Foundation Berlin-Adlershof. Employing a synchrotron facility allows for a variation of the photon energy of

2.2 CdSe nanoparticle shape evolution tuned by halogenated additives

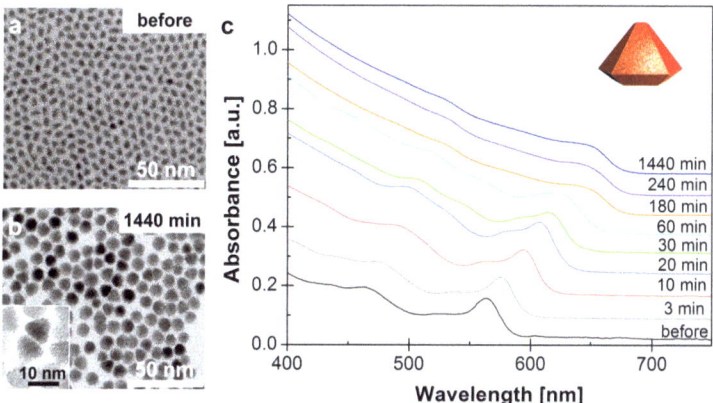

Figure 2.10: (a, b) Micrographs of samples before injection of 1-chlorooctadecane and after 24 hours of growth in presence of the additive. The inset in (b) shows enlarged pyramids with lattice pattern perpendicular to the c-axis. The UV\-Vis absorbance shifts and broadens only gradually, before it remains in place after 3 hours.

the monochromatic incident beam, which becomes important when relevant signals overlap with Auger signals[5] from other elements. In standard set-ups with Al anodes and a photon energy of 1486.6 eV, for instance, P 2p signals overlap with LMM Se Auger signals. This problem is avoided when reducing the photon energy to 620 eV, since the orbital signals shift but the Auger signals remain unchanged. The examined samples were prepared by *in situ* deposition of monolayers onto highly oriented pyrolitic graphite (HOPG) following a recipe optimised for this purpose [125]. Even though this involved longer reaction times and slightly changed amounts of ligands, the overall morphological evolution was comparable to the reactions without carbon substrates.

The three samples examined were nanorods without additive, nanoparticles prepared with bromine additive DBE and with chlorine additive DCE. Figure 2.11 displays TEM micrographs of nanoparticles from the supernatant above the HOPG substrates, XPS survey scans, high resolution P 2p signals of all three samples and high resolution data of the relevant halogen peaks. It was established that nanopyramids grow in solution and then attach to the carbon surface [16], so that the shape of the attached nanoparticles is assumed to be identical with the ones shown.

---
[5] Auger electrons are generated by ejection from outer shells through energy that is released when an energetic core electron fills the vacancy left by the electron first ejected by a photon.

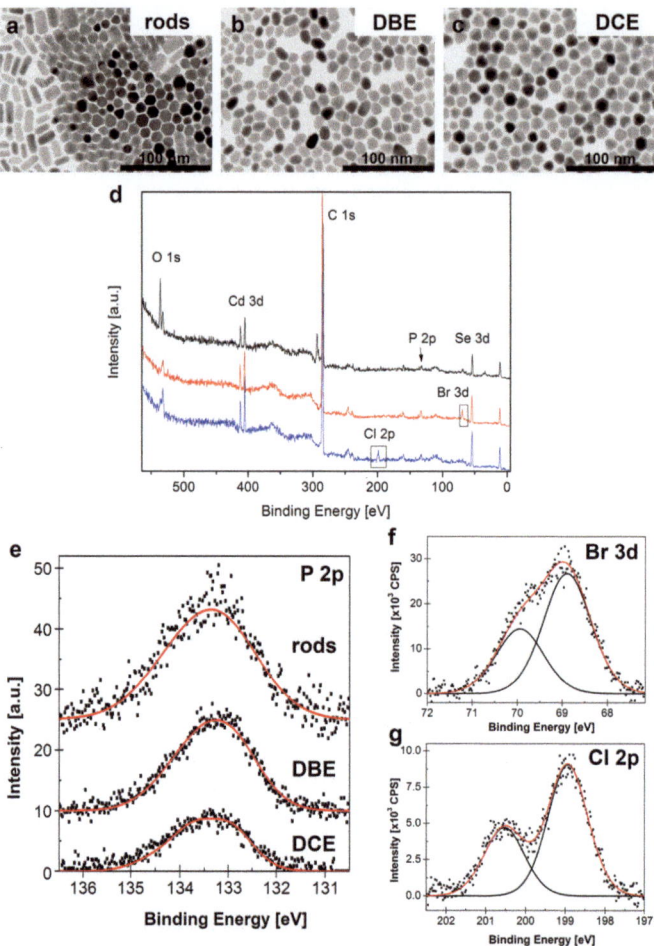

Figure 2.11: (a - c) TEM micrographs of nanoparticle aliqouts isolated from the supernatant of XPS samples *in situ* deposited on HOPG. (d) XPS survey scans of rods and pyramids prepared with DBE and DCE (top to bottom) obtained at 620 eV with an energy pass of 50 eV. (e) High resolution scans of P 2p signals for all three samples (energy pass 20 eV); the area of the peaks was normalized to the area of the respective Cd signal. (f) Br 3d orbital signal of the sample prepared with DBE and (g) Cl 2p signal of the sample obtained with DCE. In (d - g) red lines are the envelope functions of the applied fits and the fits are shown as a solid black line. In (e) the fit and envelope functions are identical.

The Cd and Se XPS signals of the samples did not show significant differences, which is due to the large bulk contribution from the nanoparticle core, which masks comparatively small changes on the surface. Judging by the widths and positions of the halogen peaks an interaction of the atoms with the nanoparticle surface can be assumed. The peak position of Cl 2p 3/2 of the sample prepared with DCE is with 198.8 eV close to the value reported for spherical Cl-capped CdSe nanoparticles and in between the ones in bulk phases of $CdCl_2$ (198.4 eV) and the more ionic $ZnCl_2$ (199.1 eV) [29, 125, 126]. This circumstance hints at an ionic interaction and supports the presumption that halide ions are the species adsorbing to the surface and interfering with the shape evolution. Similarly, the Br 3d 5/2 signal, peaking at 68.9 eV, lies in between $CdBr_2$ (68.6 eV) and $ZnBr_2$ (69.4 eV) [126]. In addition to this, the full width at half maximum (FWHM) of both signals is broader than the reference signal (Cl 2p 3/2: 1.2 eV; Br 3d 5/2: 1.3 eV; Au 4f: 0.7 eV), indicating different chemical environments for the halogens, for instance caused by the interaction with different nanoparticle facets. A comparison between the P signals, where the areas were normalized to the area of the corresponding Cd 3d signal, reveals a reduction of the relative P content in the samples from rods (0.14) to nanopyramids prepared with DBE (0.09) to nanopyramids prepared with DCE (0.05). Relating this tendency with the above shown micrographs where the nanoparticles with DCE are of pyramidal morphology while those with DBE are of a rounder morphology after the same reaction time, leads to the assumption that the shape evolution induced by halogens is accompanied by a loss of P ligands.

Iacono *et al.* revealed that CdSe nanopyramid attachment to HOPG is promoted in samples with higher Cl content [125]. There is an inverse correlation of the relative peak areas between P/Cd and Cl/Cd in rod-shaped and pyramidal samples with different Cl content. In combination with solid state nuclear magnetic resonance spectroscopy and inductively coupled-mass spectrometry this was attributed to a reduced amount of ODPA-related ligands, especially anhydrides, and incorporation of Cl into the ligand sphere, allowing for a more intimate contact between nanopyramids and substrates.

Changes of P and Cl contents during the shape evolution in nanoparticle samples prepared after the method without carbon substrate were examined by standardless total reflection X-ray fluorescence spectroscopy (TXRF) in collaboration with Mauricio D. Coderch and Ursula A. E. Fittschen from the Institute of Inorganic and Applied Chemistry of the University of Hamburg. The elemental analysis of nanoparticles and their ligand sphere is not straightforward. In the present case, the components of interest, P and Cl, are constituents of the ligand sphere and thus only a small fraction of the total amount of atoms in a sample. There is a limited number of analytical methods capable of determining Cd and Cl simultaneously under this pretext. For the determination of Cl in the relatively low concentrations found in the present samples, special systems of inductively coupled mass spectrometry (ICP-MS) or optical emission spectroscopy (ICP-OES) would be necessary. Furthermore, the mandatory dissolution of the samples prior to ICP-MS or ICP-OES analysis leads to problems with a quantitative capture of Cl after acidic digestion. Through the formation of gaseous hydrochloric acid major amounts of Cl are lost[6]. In energy dispersive X-ray spectroscopy (EDX) in combination with transmission electron microscopy (TEM) light elements are usually detected well but their low content in the samples resided in the range of the analytical error of the method.

A more convenient and applicable method for elemental analysis of nanoparticles without the need for digestion processes is TXRF. Samples may be deposited onto the sample holder as a powder or directly from solution. A small drawback is that P and Cl can be quantified but a satisfactory accuracy in their determination results in larger errors of absolute values for heavier Cd and Se due to re-absorption effects. In addition, the determined Cd values are too high, which is caused by an overlap of the Cd L-signals used for calculation by the evaluation software (*Spectra*) with background Ar. For these reasons trends but no absolute values should be taken into account in this case. Remarkably, already aliquots with rod-shaped nanoparticles taken after 10 minutes contain Cl, which confirms an influence of the halogen throughout the reaction. A temporal evolution over the course of a reaction is shown in Figure 2.12. Clear trends in the atomic ratios relative to Cd are apparent, with an increase of Cl and a decrease of P values with time and thus proceeding shape evolution. This dependence further supports a mixed coordination of Cd with P and Cl ligands and is in accordance with the relation found in reference [125]. Despite this repeatedly observed relation between P, Cl and Cd, the Se to Cd ratio changes without recognisable trend which must be due to reasons other than the morphological evolution. An explanation might be the presence of residual precursor material. Even though the purification procedure was leaned on reported protocols for CdSe nanoparticles obtained with similar components [78, 127], there might still be varying amounts of impurities in the analysed samples, calling for a further optimisation of the process for quantitative elemental analysis.

---

[6] The same applies for Se, which can partially evaporate as hydrogen selenide. A basic alternative such as the digestion after Schöninger in an arc lamp and collection of the ashes in sodium hydroxide solution would be ideal to determine the Cl content. Anyhow, the reaction mixture and nanoparticles contained several oxygen bound phosphorous compounds which can form explosive mixtures with air upon intensive heating.

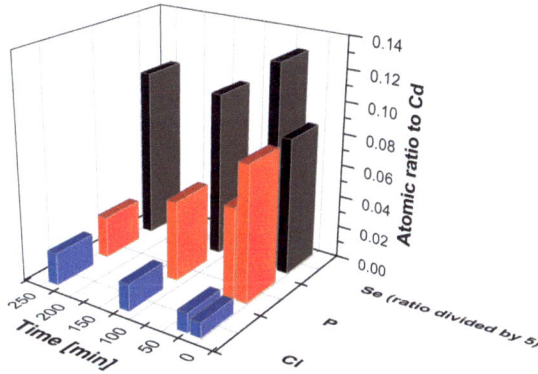

Figure 2.12: Temporal evolution of atomic Cl, P and Se to Cd ratios as measured by TXRF. The values for Se were divided by 5. Errors arising from sample preparation are Cl: 9.2%, P: 20% and Se: 6%.

On the other hand, as Anderson *et al.* formulated, "nanocrystal purity is an indefinite concept" [108]. An important question in this context is where the actual limits of a nanoparticle and its ligand sphere are set. For physical properties, the inorganic core of the particle and the ligands that fill surface traps will be decisive. For the chemical reactivity everything that sticks to the nanoparticle during common purification procedures will be important, so that excessive purification will prevent an accurate analysis. The aforementioned authors proposed that the outer layer of CdSe nanoparticles is composed of ligand coordinated Cd, which improves quantum yields and can be washed away easily, thus distorting the results. For these reasons and due to incompatibilities of the samples with many standard work-ups, the analytical description of nanoparticles seems to be a compromise between necessary purification and determination of the real sample constitution.

2.2.4 Ligand-surface interactions and the hexagonal pyramidal shape

On the first glimpse the pyramidal morphology does not seem to be energetically favourable. Nevertheless, naturally occurring cadmoselite minerals with hexagonal dipyramidal *Habitus* were found [128] and the shape is formed at a point in the reaction where a transition from kinetic to thermodynamic growth control is expected. Together with the fact that this morphology is a polyhedron, which is likely formed under thermodynamic control (see section 2.1.2), it appears reasonable that hexagonal pyramids are indeed an equilibrium shape of the wurtzite structure in the presence of halides. To clarify this point, the already mentioned experimental and theoretical studies of Wolff come to help. For wurtzite crystals with ionic or partially ionic character the contribution of cationic bonds to the surface energy is higher than that of anions. Thus, anion rich facets constituting hexagonal pyramidal morphologies with large $(000\bar{1})$ and $\{10\bar{1}\bar{1}\}$-type facets are predicted [66]. An increasing degree of ionicity determines whether the pyramids exhibit a frustum with mixed ions or if they are flat and completely anion rich. Such expected anion-rich pyramidal crystals were observed experimentally for vapour grown wurtzite materials including CdSe, while, on the contrary, cation rich surfaces with predominant (0001) base formed in solution by etching with hydrochloric acid. The latter was ascribed to an interaction between water molecules or halogen and the crystal surface. This demonstrates that the adsorption of ionic species exerts a significant impact on the surface energies and shape of crystals and that the shape of CdSe nanopyramids is most likely caused by Cd bound chloride. A slight increase of ionicity and growth control due to halide adsorption might also explain the above described preference for the wurtzite over the zinc blende phase in other reports.

To further elucidate the role of different ligands in nanorods and nanopyramids, density functional theory (DFT) calculations under aperiodic boundary conditions were carried out by Christian Klinke. Binding affinities to the dominating facets of hexagonal pyramids of possible ligands and additives were determined with the ORCA software [129]. More specifically, the LDA exchange functional, the correlation functional VWN-539 and the Ahlrichs TZV basis set were applied [130, 131]. The adsorption energies are the difference between the sum of the separate energies of the CdSe crystal and the ligand and the total energy of the combined system. The CdSe slabs visible in Figure 2.14 were kept at constant geometric parameters, whereas the ligand was free to relax. The organic ligands coordinating CdSe nanorods and pyramids are related to octadecylphosphonic acid [132], in particular double deprotonated ODPA^{2-} and deprotonated anhydrides with two or more connected molecular units [125]. Other potential ligands in the reaction are tri-n-octylphosphane and its oxide, as well as the haloalkanes with DCE as representative and released atomic halides. Earlier calculations have shown that there is little difference between adsorption energies of short and long chained ligands [96]. For this reason, calculation time could be saved by employing short alkyl chains (propylphosphonic acid, PPA,

2.2 CdSe nanoparticle shape evolution tuned by halogenated additives

instead of octadecylphosphonic acid) without changing the qualitative aspects of the results. The data obtained with facets most relevant for CdSe rods and pyramids (depicted in Figure 2.13) is listed in Table 2.2.

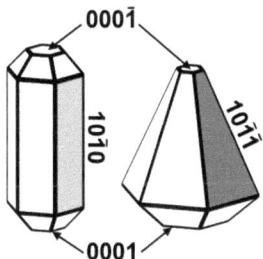

Figure 2.13: Schematic depiction of a rod and a hexagonal pyramid with distinct facets.

Table 2.2: Adsorption energies of different ligand species on dominant CdSe nanorod and -pyramid surfaces (PPA: propylphosphonic acid; TPP(O): tri-n-propylphosphine(oxide)).

Ligand	Adsorption energies [eV]			
	CdSe (0001)	SeCd (000$\bar{1}$)	side (10$\bar{1}$0)	slope CdSe (10$\bar{1}\bar{1}$)
PPA	1.86	0.97	2.28	2.15
PPA-anhydride	2.18	0.97	2.82	(6.23)*
TPP	1.76	1.04	2.08	1.81
TPPO	1.93	0.91	2.37	2.22
DCE	0.90	0.49	1.19	0.77
PPA^{2-}	11.44	6.11	9.03	12.90
PPA-anhydride^{2-}	10.40	4.40	8.26	12.57
Cl$^-$	4.39	1.80	3.45	5.49
Br$^-$	4.14	1.70	3.21	5.20
I$^-$	4.07	1.80	3.15	5.06

* Decomposes upon interaction.

Stronger adsorption energies result from the calculations for all charged X-type ligands (deprotonated phosphonic acid species, halides), while neutral L-type ligands (TOP, TOPO, protonated phosphonic acid species, DCE) interact only weakly on all facets. These findings are in agreement with previous reports on ligand-CdSe surface interactions [31, 77, 78]. All species adsorb comparatively weakly to the Se-rich (000$\bar{1}$) facet, whereas they interact more strongly with Cd-rich sites (see also [77, 96, 133]).

CHAPTER 2. HALOGEN INDUCED SHAPE CONTROL OF CDSE NANOPARTICLES

For the L-type ligands, the highest adsorption energy is found for the non-polar ($10\bar{1}0$) side facet terminated with both Cd and Se atoms. It is followed by the values for the Cd-rich ($10\bar{1}\bar{1}$) slope and (0001) bottom facets. The least favourable interaction occurs with the Se-rich ($000\bar{1}$) facet. The preference of these ligands for the side facet confirms earlier calculations which were utilised to explain the growth of nanorods [77].

In comparison to the X-type ligands, however, the binding affinity is weak. With the charged ligands ODPA^{2-} and ODPA-anhydride^{2-} and halides the adsorption energy follows a different order. Here, the interaction on the Cd-rich facets is energetically most favourable and decreases from the ($10\bar{1}\bar{1}$) slope to the (0001) bottom, ($10\bar{1}0$) side and ($000\bar{1}$) top facets. The ligands most strongly interact with the sloped ($10\bar{1}\bar{1}$) facet because this is the roughest one terminated by Cd sites. From the slab models depicted in Figure 2.14, it can be seen that Cd is compactly coordinated by 3 Se atoms on the (0001) facet (bottom). In the mixed ($10\bar{1}0$) side facet, Cd is also coordinated by 3 Se atoms albeit in a much more corrugated fashion. Importantly, in the ($10\bar{1}\bar{1}$) slope facet, Cd atoms have two dangling bonds instead of one, since they are only coordinated by 2 Se atoms. This leads to a higher tendency of ligand coordination and the high adsorption energies for charged ligand species, especially the dominant ligand ODPA^{2-}. Another remarkable result is that protonated (neutral) phosphonic acid anhydrides were unstable on the sloped facet and even decomposed. Halides on the contrary prefer adsorption to this facet. The energies decrease with the trend of sinking ionicity: $Cl^- > Br^- > I^-$. Taking into account the tendency towards the pyramidal shape observed in the reactions, it occurs reasonable that the ionicity of the surface bonds is the driving force, in agreement with the Wolff model. In relation with the other ligands, halides with their single charge exhibit adsorption energies between the L-type ligands and the double-deprotonated ODPA species. With the observed adsorption energies and the contributions of Wolff it is feasible to claim that the pyramidal shape is energetically favourable under conditions where ligand-surface interactions and rearrangement of surface atoms determine the growth rates of different crystal facets. Such conditions are fulfilled under thermodynamically controlled ripening.

The special role of Cl^- might be explained by a mixed coordination of Cd-rich slope facets by Cl^- and ODPA^{2-}. Due to the steric demand of deprotonated phosphonic acid ligands there is space enough for the adsorption of small halides to additional coordination sites which even bulkier anhydrides cannot reach. Such a mixed coordination complies with reported trap filling in PbS samples by halides, where the latter sit in between or even partially substitute oleic acid ligands [27, 28].

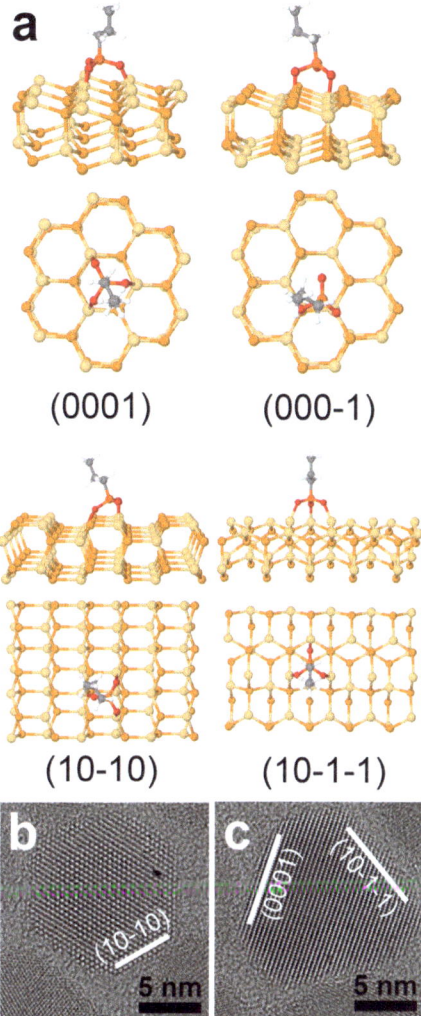

Figure 2.14: (a) Side and top-view of PPA^{2-} adsorbed to different CdSe facets and HR-TEM images of pyramidal CdSe nanoparticles, (b) top and (c) side-view with indicated facets. Atoms in beige: Cd, orange: Se. Reprinted with permission from [124]. Copyright 2014 American Chemical Society.

CHAPTER 2. HALOGEN INDUCED SHAPE CONTROL OF CDSE NANOPARTICLES

2.2.5 A closer look at the shape evolution process

The presented results lead to the assumption that halogenated additives affect the morphological evolution in consecutive kinetic and thermodynamic growth stages in different ways. This can be understood by bringing together experimental and theoretical discoveries. In Figure 2.15 a LaMer-type plot of the reaction with the two most important influences of the additives, the increase of rod size and the tendency to reshaping, is shown.

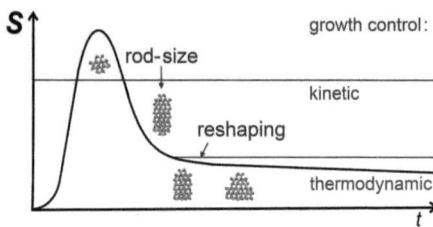

Figure 2.15: Morphological evolution of CdSe nanoparticles to pyramids under influence of halogenated additives in connection with different modes of growth control. Adapted with permission from [124]. Copyright 2014 American Chemical Society.

Rod growth

Both in reactions with and without additives, the growth rate in the direction of the c-axis is the highest during the first stage of growth, which can be related with the low adsorption energies of ligands to the $(000\bar{1})$ facet (see Table 2.14). The other facets are tightly capped by a mixture of ODPA^{2-} and deprotonated anhydrides which can bridge adjacent Cd atoms and isolate the surface from the solution. /par The effect of chlorine containing additives with ratios to Cd of up to 0.7 in this stage of growth is an increase of precursor and/or solute solubility and thus of the rod size. The extent of this is more visible with larger amounts and additives with sterically more accessible carbon centres bound to chlorine atoms. This tendency, together with the possibility of inducing the shape transformation with chloride compounds and the signs for ionic interactions between Cl and the nanoparticle surface, strongly suggests that released Cl$^-$ X-type ligands are the responsible ingredient.

At a DCE/Cd ratio of 1.0 and with 1,2-dibromoethane and 1,2-diiodoethane at DXE/Cd: 0.7 the additive does not only increase the nanoparticle size but also affects the roughness of the side facets. Zigzag faceting is observed in parallel to the c-axis (section 2.2.2.1). Similar nanorod shapes were obtained in experiments with ligand mixtures of hexylphosphonic acid

34

and TOPO, in which higher contents of the prior lead to the formation of arrow shapes [82]. Such shapes contain polar, presumably Cd-rich facets, which insinuates that shorter phosphonic acids, or in this work halogen compounds, compete with octadecylphoshonic acid for surface coordination and that there is a threshold above which the shorter ligand induces the expression of Cd-rich diagonal facets already during the first stage of growth. An explanation for the stronger impact in this direction by the heavier halides may be linked to their leaving group ability. It increases in the line $Cl^- < Br^- < I^-$. For this reason, Br^- and I^- are more readily released from haloalkanes than Cl^- and present at higher concentrations in the beginning of growth. This way, they can affect the evolution of the side facets in a similar way as Cl^- at a higher additive concentration. In TXRF measurements with aliquots taken after 30 minutes the measured atomic halogen/Cd ratio was indeed slightly higher with DIE than with DBE and DCE (DIE: 0.021, DBE: 0.018, DCE: 0.012).

Ripening to pyramids

The transition to ripening and evolution of the pyramidal shape is a smooth process happening under interaction with solute in the liquid phase and is accompanied by dynamic ligand adsorption and desorption as well as rearrangement of atoms. These conditions are different to studies reporting a post-synthetic formation of pyramidal CdSe nanoparticles by etching of Se-sites of CdSe spheres in basic solutions containing a Cd-ligand or other shape transformations by photoetching in presence of trichloromethane [134, 135, 136]. With proceeding reaction time the growth rates change and the formerly well protected side facets grow fastest while stable sloped facets develop. The reshaping process begins with etching on one of the tips which was identified as the $(000\bar{1})$ terminated one by TEM tomography studies [106]. A suitable explanation would be the removal of weakly capped Se by chloride [137]. However, the circumstance is more complex because the presence of water, which is a side product of the complex formation between Cd and ODPA [138], also plays a role. It was shown to promote the attachment of CdSe nanopyramids to carbon allotropes [16] which in turn depends on the Cl-content of the particles [125]. By implication this means that water promotes the incorporation of Cl into the ligand sphere. Potential effects of water in the reaction are (i) increasing the conversion of TOPSe [116], (ii) activating Cd-ODPA complexes by protonation [139], as well as (iii) supporting the desorption of Se^{2-} and phosphonates from the surface by protonation (neutral ODPA binds less strongly than deprotonated $ODPA^{2-}$, see Table 2.2) and (iv) increasing the polarity in the reaction mixture to better dissolve ionic species (TOP/TOPO).

CHAPTER 2. HALOGEN INDUCED SHAPE CONTROL OF CDSE NANOPARTICLES

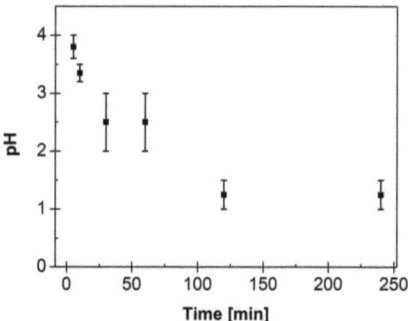

Figure 2.16: Plot of the pH of aqueous phases after extraction of organic reaction aliquots against time. The error bars correspond to the accuracy of the pH tests.

Changes in the amount of free protons during the reaction were investigated by extracting defined volumes of aliquots with water and determination of the pH of the aqueous phases. A decrease of the values with time is visible in Figure 2.16. The origin of this might either be an elimination of protons from the additives together with the halogens or the reaction of excess octadecylphosphonic acid with the nanoparticles leading to partial dissolution of CdSe under formation of hydrogen selenide [29]. With the amount of free protons and ratios of Cl to Cd found by TXRF increasing together with the evolution of the pyramidal shape, parallels to Wolff's etching of CdSe by hydrochloric acid occur. Under these conditions protonation of octadecylphoshonate (ODPA^{2-}) and its anhydrides, tightly bound to the flat side facets of rods, might be protonated. This would substantially reduce their adsorption energy and eventually lead to their detachment from the surface. Once this happens, the surface atoms start to rearrange or be dissolved and the new shape begins to evolve. Halide ions may then be able to occupy coordination sites of forming facets. This way, they would prevent the adsorption of sterically demanding anhydrides, which decompose on the $(10\bar{1}\bar{1})$ facet, or phosphonates at these positions. Additionally, they may fill sites that are inaccessible to bulkier ligands. In the row of the halides Cl$^-$ is the smallest and most strongly Cd-binding ion. For this reason, it is able to diffuse faster than the others and may thus win over the phosphonate ligand in more cases so that it affects the growth direction more strongly.

Another explanation would be that the solubility of surface atoms and the rearrangement depend on the type of halogen. Saruyama *et al.* suggested that a formation of pencil-shaped hexagonal Cd-chalcogenide particles from zinc blende seeds in a mixture of different ligands and an organic ammonium chloride salt was driven by Ostwald ripening under the dissolution of Cd with Cl$^-$ and the help of other ligands [140]. In the current reaction it is possible that there is an influence of Ostwald ripening but the size dispersion remains below the more than 20% expected for purely ripened samples. Even more importantly, in the sample with a low DCE/Cd ratio a minority of nanoparticles had developed a pyramidal morphology, irrespective of size. A pure Ostwald ripening process would favour pyramidal growth only for large nanoparticles. In consequence, there might be an overlap of reshaping to pyramids and Ostwald ripening.

2.3 Conclusions

Halogen compounds severely impact the shape evolution in the hot-injection synthesis of CdSe nanorods. In the presented method haloalkanes influence both the kinetic and thermodynamic growth regime of the reaction leading to hexagonal pyramidal equilibrium shapes which form through ripening of rod-shaped intermediates. By experimental, analytical and theoretical approaches the reasons for the formation of this particular morphology and opportunities to modify the method in order to tune the shape of nanoparticles were examined.

Similarities in the role of haloalkanes and ionic halide compounds were observed, in which structural properties and the added amount of haloalkanes influence the growth of the nanoparticles and the rate of their morphological evolution. The reactivity of the additives seems to be related to their ability to release halides and thus generate X-type ligands *in situ*. With higher relative availability of halides in the kinetic growth regime the size of nanorods was increased, which is attributed to a higher solubility of the precursors and/or solute in the reaction medium. At a 1,2-dichloroethane to Cd molar ratio of 1.0 or with analogous bromine and iodine compounds at a ratio of 0.7 zigzag faceting of the anisotropic nanoparticles occurred. This indicates a competition between halides and the original octadecylphosphonic acid related ligands for determination of the growth direction. If the concentration of additive was comparatively low (molar ratio 0.3 for DCE), the nanoparticles were affected homogeneously leading to a distribution of sizes and shapes. A relation between the size of nanorod intermediates and the final nanopyramids was observed. This opens the possibility to tune the size of rods by variation of the additive or the moment when it is injected and produce nanopyramids with different dimensions.

CHAPTER 2. HALOGEN INDUCED SHAPE CONTROL OF CDSE NANOPARTICLES

A comparison with different halogens in 1,2-dihaloethane revealed that the tendency to affect the nanoparticles in the kinetic growth stage falls from I^- to Cl^- whereas the influence on the extent and rate of reshaping to pyramids follows the opposite direction. From this it was concluded that the evolution of the pyramidal shape is not only related to the amount of halide present but also to their interaction with the nanoparticle surface.

By synchrotron XPS and TXRF a loss of phosphonate ligands and an increase of halogen content from rod-shaped to pyramidal samples were registered. This was set into a theoretical context by combination with density functional calculations and a model that predicts crystal shapes by their periodic structure and contributions of dangling bonds on the surface. As a result, the evolution towards a pyramidal shape could be explained as originating from an energetically favourable mixed coordination by phosphonate and halide ligands to sloped $(10\bar{1}\bar{1})$ wurtzite facets in the thermodynamically controlled ripening stage of growth. The peculiarity of this facet is the higher number of dangling bonds of Cd surface atoms and rough structure. An increase of free protons in the reaction mixture and of Cl^- with proceeding ripening illuminate parallels to the formation of hexagonal pyramids in macroscopic acidic etching [65]. The difference is that the particles continue to grow toward an equilibrium shape because of remaining solute in the liquid phase.

The findings provide an explanation for the formation of morphologies with distinct vertical facets observed in wurtzite structured II-VI semiconductor nanoparticles and are expected to be transferable to other materials. The *in situ* generation of surface active species is not only a factor to keep in mind when unforeseen shapes form during the synthesis but may also be applied as a versatile tool for the design of new nanoparticle morphologies.

3 Metal-semiconductor hybrid nanoparticles

Colloidal hybrid nanoparticles belong to an emerging class of multi component materials that offer vast possibilities of creating structures for advanced applications in photocatalysis [1, 141, 142], in optoelectronics [143] and additionally in biology and medicine [144, 145]. Hybrid formation, together with ion exchange and galvanic replacement techniques, presents a significant step towards the combination of nearly all materials available in purpose-designed morphologies [146, 147, 148]. Materials with different physico-chemical properties may be joined to complement each other and exhibit new collective properties. Among these are effects occurring by coupling of the optical responses as described below or second harmonic generation [149]. Alternatively, multifunctional nanoparticles maintaining the original features may form, which for example exhibit magnetism and fluorescence at the same time (Co/CdSe core-shell structures [150]). During recent years, a large variety of structures and compositions has been developed. A detailed overview of material combinations, preparation methods and properties is provided by reviews [8, 46, 151, 152]. Synthetic aspects and mechanisms are in the focus of reviews [153, 154, 155]. Metal-semiconductor combinations are especially attractive hybrid materials owing to their electronic interactions [152]. On the one hand, grown-on metallic contacts improve electrical transport through semiconductor structures [13] while the interfacial contact on the other facilitates exciton/charge separation, a valuable property for the strongly emerging field of (photo)catalysis [1, 142, 156]. In the following, the focus will lie on such metal-semiconductor combinations, in particular on metal-cadmium chalcogenide hybrid nanoparticles.

Thorough knowledge and control of the configuration of a material is a prerequisite for studies towards application. Changes in the material caused by the examination may lead to unexplainable or unreproducible results. To avoid such problems it is important to understand the mechanisms and roles of the components active in the formation process of hybrid nanoparticles. This additionally helps to tune the morphology by deliberately varying the conditions.

In this chapter, the focus of the experimental work lies on the colloidal seeded-growth preparation of hybrid nanoparticles from hexagonal pyramidal CdSe nanoparticles treated in Chapter 2. Major points of investigation are occurring instabilities of deposited metal domains and conditions leading to oligomeric structures with several spherical metal domains decorating the semiconductor. The peculiar morphology of the pyramidal nanoparticles resulted in a deposition behaviour different from the widely examined rod-shaped seeds. Four metal-CdSe combinations were studied to find promising, stable candidates for electrical studies. Before these results will be presented and discussed, properties and aspects of hybrid nanoparticle synthesis including deposition and ion exchange phenomena will be introduced.

3.1 Properties and synthesis of metal-semiconductor hybrid nanoparticles

Metallic domains are capable of electronic interaction with semiconductors across a shared interface. Furthermore, they may be utilised for functionalisation with organic molecules, to weld semiconductor nanoparticles [157, 158], or to grow a third material on top [159]. Within the class of metal-semiconductor materials a variety of morphologies is obtained. It includes rod-based matchstick and dumbbell structures (Au-CdSe [10]; Au-CdS [160, 161]; Ag-CdSe [162]; Pt-, PtNi- and PtCo-CdS [163], Pt-CdS[164]) as well as metal-core/semiconductor-shell structures (Co/CdSe [150]; Au/PbS [165]; Au, Pt/II-VI semiconductors [166]) and hetero-dimers (Au-CoPt$_3$ [167]; Au-CdSe, Au-PbSe [168]). Additionally, networks (Au-CdSe [158], [169], [170]) and spherical, one-dimensional or multigonal semiconductor seeds with more than two metal domains (Au-PbS [171, 172]; Au-CdSe [173, 174, 175, 176]; Au-CdS [175]; Au-PbSe [177]; Au/Ag$_2$S-CdSe/CdS [178]; Pt-, PtCo- and PtNi-CdSe [174]) were reported. The morphology, size and number of metallic domains strongly influence the properties of the nanostructure.

3.1.1 Properties of metal-semiconductor hybrid nanoparticles

The most relevant features of metal-semiconductor nanoparticles are based on electronic interactions of the components. These are strongly influenced by the relative positions of the semiconductor bands and the work function of the metal. Figure 3.1 contains a schematic representation of the relative positions of electronic levels of CdSe in bulk and nano dimensions and the bulk values for the metals Ag, Au, Pd and Pt. Positions of band edges in nanoscale materials differ widely throughout the literature, which is partially caused by the employment of different methods for their determination. The position of the topmost level of the valence band, for example, is not only size dependent but also influenced by the head group of coordinating ligands [69]. The values in Figure 3.1 were calculated for CdSe nanospheres with dimensions similar to those of CdSe nanopyramids as described in Chapter 2. Work functions of metals depend on the crystal facet. This makes clear that the relative positions in the macroscopic state may only serve as a rough orientation for hybrid nanoparticle systems. When the two components are in the ground state and form a junction, electrons begin to flow in order to equilibrate the Fermi levels. This may cause the bands of the semiconductor to bend. Such band bending may also affect optical and catalytic properties [179, 180] but it is a matter of discussion regarding the results reported in this context. For this reason it will be treated phenomenologically. Anyhow, it has significant consequences for the electrical properties of hybrid nanoparticles and their solid state assemblies and will thus be addressed in Chapter 4.

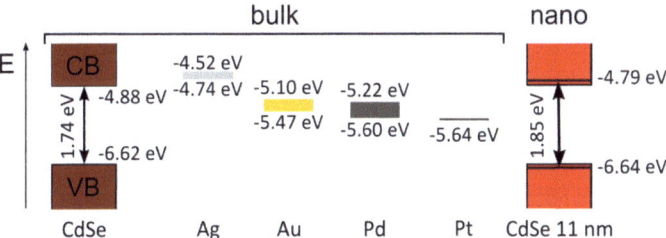

Figure 3.1: Schematic depiction of electronic bands of CdSe bulk material, ranges of bulk work functions reported for different noble metals and the lowest electronic transition for nanoparticles with a diameter of 11 nm. The bulk values of the valence band (VB) edge position (determined by the photoelectric effect) and the band gap of CdSe are cited from [181] and [59] (CB is the conduction band). The different metal work functions, listed in [182] and supplemented by [183], depend on the crystal facet and were obtained through the photoelectric effect. The band gap shown for nanoparticles is the converted absorption wavelength (670 nm) of spheres with a diameter of 11 nm [184]; the level positions were calculated with bulk constants[7].

---

[7] The positions of the energetic levels were determined from the confinement energy $E_{conf.} = h^2/(8r^2 m_{eff.})$ of electron and hole [57]. Apart from the Planck constant h, effective masses $m_{eff.}$ of $0.13 \times m_0$ for the electron in the conduction band and $0.45 \times m_0$ for the hole in the valence band were inserted [185].

CHAPTER 3. METAL-SEMICONDUCTOR HYBRID NANOPARTICLES

For solution based applications the transfer of excited electrons becomes relevant [186]. If the metallic compound exhibits plasmon resonance, plasmon-exciton coupling or so-called *plexitonic interactions* may occur [187], which are exploited for tailoring light-matter interactions [188]. The nature of these interactions depends strongly on the size and shape of the metallic domain, since this determines whether the metal is capable of exhibiting a plasmon resonance. In contrast to exciton formation in semiconductor nanoparticles, visible light excitation of some metallic nanoparticles does not result in antipodal charge carriers but in collective oscillations of electrons, referred to as surface plasmon resonance. Position and shape of the resonance absorption curve depend on the material, the size and the shape of the nanoparticle along with the dielectric constant of the surrounding medium [189, 190, 191, 192, 193]. The onset of plasmon resonance phenomena with Au nanoparticle size, for instance, is marked by a transition from localised electronic states in Au nanoclusters of sizes below 2 nm to band-like electronic structures in bigger crystals [189, 193]. Apart from Au, Ag and Cu exhibit plasmon resonances with strong absorption maxima in the visible part of light. The other noble metals show plasmon absorption further towards the ultraviolet region with little featured curves caused by lifetime broadening through various decay processes [190, 194]. Interactions between semiconductor and metal nanostructures separated by defined distances, for instance by layers of polyelectrolyte or silicon dioxide [195, 196, 197], tend to enhance photoluminescence properties of the semiconductor component. Direct interfacial contact often leads to bleaching of the surface plasmon resonances and suppression of photoluminescence [198]. In optical absorption and emission, orbital mixing across the interface and charge transfer manifest themselves in a broadening of the absorption maxima and photoluminescence quenching on the part of the semiconductor [10, 160, 171, 199]. The intensity of these effects scales with the size of the metal domains (compare [180]). With Cd-chalcogenides plasmonic interactions only appeared with large metal nanoparticles if direct contact existed [160, 161, 200]. The experimental findings fit with theoretical predictions by Govorov *et al.*, where the semiconductor component is treated as an emitter that may be amplified or damped by surrounding metal nanoparticles [201]. A recent report demonstrated photoluminescence enhancement for oligomeric Au-Cd chalcogenide structures [202]. The authors explained this circumstance by a reduction in the number of trap states and collective plexcitonic interactions in the nanostructures.

Depending on the relative positions of the energy levels and the wavelength of the incoming irradiation, a flow of excited electrons can occur in both directions [203]. Electrons

3.1 Properties and synthesis of metal-semiconductor hybrid nanoparticles

generated in the semiconductor component are transferred to a metallic domain if the metal work function lies below the conduction band edge. If the hole remains in the semiconductor or is transferred to a scavenger, charge separation occurs. A transfer of electrons to the metal is advantageous for optoelectronic applications [3, 204, 205] or subsequent reduction of chemical species in solution based photocatalytic approaches [1, 11, 156]. The efficiency of the process is correlated with the type of metal in the hybrid nanoparticles. In ZnO and TiO_2 hybrid nanoparticles, for example, charge retention occurred with Au and Ag domains in water while Pt quickly transferred the electrons to the surrounding solution [206, 207, 208]. A similar trend was observed with Au-CdSe and Pt-CdSe nanodumbbells in organic solution [199, 209]. The difference was attributed to Fermi level equilibration between the components with Au in both cases, resulting in a reduced transfer. While metal oxides rely on UV-irradiation, cadmium chalcogenides can be employed in visible light photocatalysis due to the favourable position of their size dependent band gap [151]. Au tipped CdSe rods were shown to promote charge separation and retention [11, 204]. Compared with pure CdSe nanorods, the dumbbell structures exhibited a more than threefold increase in photocatalytic methylene blue reduction. Pt decorated CdS, CdSe and CdSe/CdS rods and nets were intensively studied in terms of photocatalysis with a focus on hydrogen production [210, 211, 212, 213, 214, 215]. Small clusters proved to be catalytically more active than larger particles. While charge transfer occurred faster in nanodumbbells than in nanorods with one domain, the latter were more efficient in catalysis in the combination Pt-CdSe [209]. An emerging class of hybrid photocatalysts are those with plasmonic metal domains [216]. An observed increase of the photocatalytic activity in the water splitting reaction raises hopes for more efficient future systems [186, 216, 217].

Similar aspirations are pursued in photovoltaics, where plasmonic enhancement of the absorption could be demonstrated for metal coupled semiconductor films [218, 219]. Nevertheless, the nature of the interactions is under discussion with resonant excitation and simple scattering of light by the metal both being reasonable explanations.

3.1.2 Synthesis of metal-semiconductor hybrid nanostructures

A commonly applied and straightforward approach for the preparation of homogeneous hybrid nanostructures is the seeded growth method in colloidal solution. With an increasingly sophisticated control over single component nanoparticles the complexity of hybrid colloids has reached a level where current methods allow for the tuning of shape and size of both components [152]. Even though the deposition of a metal onto the semiconductor from precursor compounds is the more common case, depositions of semiconductor material onto metals are also reported [154, 165, 220]. Nevertheless, the semiconductor component will be assumed to act as the seed material in the following remarks.

In some cases, the formation of solid metal domains on the surface competes with ion exchange processes, especially of cations. For this reason, a short introduction into the principles of both heterodeposition and ion exchange will be given.

3.1.2.1 Deposition of metallic domains

Decisive parameters affecting the formation of heterostructures and their morphology are, on the one hand, related to the inherent properties of the materials. These are their crystal structure and corresponding lattice mismatch as well as the miscibility of the two components. On the other hand, surface related properties play a role. Here, the nature of the facets terminating the crystals and their reactivity towards deposition are important. The latter is increased for sites with low ligand coverage or surface defects. Epitaxial growth is a goal in heterostructure formation, as the growing component takes up the orientation of the seed structure and interface defects remain low. This is a prerequisite for effective and predictable interaction across the interface. There are three major models originally describing the modes of growth at a vapour-solid interface [221]. If the strength of adsorption between the seed layer and the components of the secondary material is strong and the lattice mismatch is small, two-dimensional layers form in a *Frank-van-der-Merwe* process. The corresponding change of the total Gibbs free surface energy ΔG_{surf}

$$\Delta G_{surf} = \gamma_1 - \gamma_2 + \gamma_{1,2} \tag{3.1}$$

is positive [222, 154]. In Equation 3.1 γ_1 and γ_2 are the surface energies of the materials and $\gamma_{1,2}$ is the interfacial energy. A high surface energy of the secondary material and/or a weak interaction caused by a large lattice mismatch (>10% [221]), on the contrary, lead to a negative ΔG_{surf} and the growth of three-dimensional islands and a minimum interface (*Vollmer-Weber* mode). In intermediate regimes, a combination of monolayer formation and subsequent island growth (*Stranski-Krastanov* mode) occurs.

If transferred to nanoparticles, one extreme case would be the formation of core-shell structures with the secondary material forming a layer around the first. The other one would be the formation of oligomeric particles with several domains of the metal decorating the semiconductor. The intermediate regime could lead to the formation of metastable amorphous layers, which may transform to segregated domains upon heating [153].

3.1 Properties and synthesis of metal-semiconductor hybrid nanoparticles

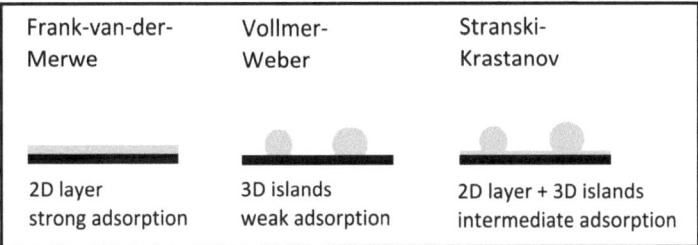

Figure 3.2: Modes of epitaxial heterodeposition [221].

An overview of strategies to prepare oligomeric nanoparticles from spherical seeds is shown in Figure 3.3. The direct deposition in (a) can be facilitated by activating the surface with a layer of the second component or one that works as a sacrificial material to reduce metal precursors [223]. The deposition in biphasic systems allows for the fabrication of Janus-type nanoparticles from precursors insoluble in one of the phases [224]. Such structures may be hydrophilic on one side and hydrophobic on the other, which may be employed in selective dual functionalisation and the formation of larger assemblies. Another method to extend the accessible configurations is the post-synthetic welding of hybrid nanoparticles. This is possible by heating, as indicated in Figure 3.3, or by addition of compounds such as sulphur or iodine that remove surfactants from the metal which results in coalescence of the nanoparticles.

With non-spherical seeds the morphology of the products will be strongly determined by the facet reactivity, as metals often deposit on the most accessible sites, for example the tips of rod- or tetrapod-shaped seeds [173]. An additional aspect that influences the morphology of hybrid nanoparticles is the mechanism of metal precursor reduction. The very first step of this process is under discussion. Some authors argue that crystalline metal domains form after precursor molecules adsorb to the surface and are reduced by anions on the seed [10, 11, 225]. Others assume an attachment and growth of metal clusters formed in solution by reduction with added amines or other mild reducing agents [154, 168]. The electrons necessary for metal domain growth can be directly provided by surface anions or by photoexcitation of the semiconductor. For anisotropic semiconductors with an intrinsic dipole moment, such as wurtzite structured nanorods, a light induced mechanism in the presence of a hole scavenger may lead to selective deposition of Au and Ag on only one tip (Ag-ZnO [9], Au-CdS [160]) or CdSe-rich sites in $CdS_{0.4}Se_{0.6}$ nanoparticles (Pd [226]). With Pt only non-selective photodeposition has been reported [227, 226]. Site specific deposition was rather observed in thermally controlled reactions without irradiation (Pt-CdS [163]).

CHAPTER 3. METAL-SEMICONDUCTOR HYBRID NANOPARTICLES

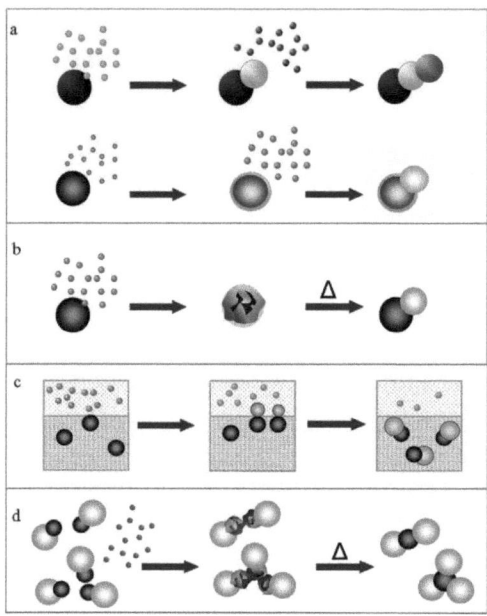

Figure 3.3: Mechanisms of oligomer formation. (a) Direct heterogeneous nucleation onto seeds without and with (lower part) activation of the surface, (b) formation of an amorphous layer and thermally activated coalescence-crystallization, (c) Janus particle formation at a liquid/liquid interface and (d) welding of hybrid nanoparticles. Reproduced from reference [153] with permission, copyright Wiley-VCH Verlag GmbH & Co. KGaA.

3.1 Properties and synthesis of metal-semiconductor hybrid nanoparticles

Several factors other than the lattice mismatch may influence the dark deposition under the pretext of a reduction-oxidation reaction between the seed surface and the metal precursor. They can be attributed to properties of the seed and of the metal precursor, as well as the reaction conditions:

- seed
  - topology of the seeds (types of terminating facets and proportion of dangling bonds)
  - reduction potential of the anion
- metal precursor
  - oxidation state/reducing power of ligands
  - concentration
  - intermetallic interactions, autocatalytic reduction
  - complex interactions with ligands
- reaction conditions
  - temperature
  - time
  - atmosphere.

In studies about metal-Cd-chalcogenide hybrid formation a frequently employed seed morphology is the rod-shape. With their weakly capped tips and comparatively strongly protected side facets (see Chapter 2), they allow for the selective deposition of one or two metal dots, depending on an increasing metal to nanoparticle ratio as in Pt-CdS [163]. With CdSe and CdSe/CdS, ripening to a one-sided structure was observed with higher amounts of Au which was explained by electrochemical Ostwald ripening of the smaller dot [228, 161]. Deposition at defect sites along the side facets of rods strongly depends on the ligand sphere with bulkier or shorter ligands rendering them more accessible than aligned primary ligands [161]. In CdS but not in CdSe the deposition behaviour depends on the presence or absence of oxygen in the reaction sphere [160].

Reduction potentials of the components may affect the kinetics of the deposition reaction [160, 229]. As shown by O'Sullivan and co-workers [230, 225], variations in the Au precursor oxidation state or the chalcogen (E) in Cd(E) nanorods allowed for rapid but controllable metal dot formation on the tips. In the row of the chalcogenides, Te(-II) is the most reactive anion followed by Se(-II) and S(-II). By utilising a precursor containing Au(I) instead of the often employed Au(III), defined Au dots may even be gained with CdTe which otherwise yields almost uncontrollable hybrid structures at room temperature.

CHAPTER 3. METAL-SEMICONDUCTOR HYBRID NANOPARTICLES

Table 3.1: Standard reduction potentials $E°$ of Cd^{2+}, Se^0 and different metal species in water[8].

Reaction	$E°$ [V]	Reaction	$E°$ [V]
$Cd^{2+} + 2e \rightleftharpoons Cd$	-0.403	$Pt^{2+} + 2e \rightleftharpoons Pt$	1.18
$Se + 2e \rightleftharpoons Se^{2-}$	0.924	$Au^{3+} + 2e \rightleftharpoons Au^+$	1.40
$Ag^+ + e \rightleftharpoons Ag$	0.800	$Au^{3+} + 3e \rightleftharpoons Au$	1.498
$Pd^{2+} + 2e \rightleftharpoons Pd$	0.951	$Au^+ + e \rightleftharpoons Au$	1.692

Reduction potentials of ionic species relevant for the present work are listed in Table 3.1. From these values, redox reactions between Se and all metal ions apart from silver are feasible. The latter may be deposited by the aid of an additional reducing agent.

The lattice mismatch between wurtzite CdSe nanoparticles (a: 4.30 Å, c: 7.01 Å [60]) and the four cubic metals is almost equal. Lattice constants a of the metals are Pd: 3.89 Å [233], Ag: 4.09 Å [234], Au: 4.08 Å [235] and Pt: 3.97 Å [236]. Based on these values, similar deposition behaviour with a formation of islands is expected in accordance with the above referred known structures. Nevertheless, there is another process which plays a role for the interaction of semiconductor seeds and metal precursors.

3.1.2.2 Ion exchange

Depending on the components and conditions a reaction pathway deviating from the formation of metallic deposits is possible, which allows for the exchange of components of a semiconductor for new elements. While maintaining the lattice structure, both cations and anions may be substituted by ion exchange processes to gain otherwise inaccessible phases and metastable structures [237]. An example for this is the preparation of zinc blend CdSe rods from PbSe [238]. The technique had been applied to bulk materials before but nanoparticles opened a new chapter for it, owing to their large surface-to-volume ratio and fast kinetics in the colloidal solution [239]. Three reviews highlight the achievements and in combination provide a profound basis for the understanding of ion exchange processes, with special regard to the more feasible cation substitution [148, 240, 241].

---

[8] This data is accepted in literature [231], even though the origin of some of it is not proven [232].

3.1 Properties and synthesis of metal-semiconductor hybrid nanoparticles

Important factors governing ion exchange are the:

- contraction of the unit cell volume $\Delta V/V$

- difference in free energy of formation of starting material and product

- solubility of the ions in the reaction solution

- chemical hardness η_A of the ions

- mobility of the ions in the host lattice/ ionic radius.

Ion exchange in nanoparticles is a phenomenon mainly controlled by kinetics but also relies on thermodynamic principles as stated in reference [241]. It is achieved more easily with smaller positive metal ions and only few cases of anion exchange are reported [242, 243, 244]. The low mobility of anions is the reason why their lattice is able to remain in position while the cations diffuse in and out of the structure resulting in topotactic changes. A prerequisite for preservation of the lattice and morphology from CdSe to Ag_2Se, for example, is that the nanoparticle is large enough [245]. A full ion exchange proceeds from the outside to the inside, with a mixed reaction zone of several nanometres between the two phases. A nanoparticle with smaller dimensions may thus undergo a thermodynamically driven rearrangement and even alter its morphology from rod-shaped to spherical.

Similar to their influence on the solubility of solute in nanoparticle crystallisation, ligands play a key role in exchange reactions by stabilising the ions in solution. By choosing their chemical hardness in correspondence to the original cation in the crystal, they may promote its solvation. This is important when ions with different oxidation states are to change places. Chemically harder ligands preferably bind to harder cations with higher positive charge, a principle that is applied for differential solvation when bivalent ions are exchanged for monovalent ones [240, 246, 247].

Table 3.2 provides an overview of cationic radii, chemical hardness and commonly observed selenide phases of metal ions employed in this work, in addition to those of the CdSe starting material. All of the listed metals are comparatively soft acids and, except for Au species, were reported to exchange with Cd in CdSe nanoparticles [245, 248]. Remarkably, the reaction with Pt resulted in substantially smaller products that might also have formed by dissolution and nucleation processes [248].

CHAPTER 3. METAL-SEMICONDUCTOR HYBRID NANOPARTICLES

Table 3.2: Ionic radii I_r, acid hardness η_A, the formula of the most relevant selenides Me_xSe_y and their crystal structure(s) of the starting material and different metal cations.

	Cd(II)	Pd(II)	Ag(I)	Pt(II)	Au(I)	Au(III)
I_r [pm]^a	95	86	115	80	137	85
η_A ^{b}	10.3	6.8	6.9		5.7	
Me_xSe_y	CdSe	PdSe / PdSe$_2$	Ag$_2$Se	Pt$_5$Se$_4$	Au(I)Au(III)Se^c	
structure	hex.^{d}	tetr.^{e} / tetr./orthorhomb.^{f}	cub./	tetr.^{g}	monocl.^{h}	monocl.^{c}

^{a} values in oxides with coordination number 6 and a radius $r(O^{2-}) = 140\,\text{pm}$ [249]; ^{b} [246]; ^{c} [250]; ^{d} [60]; ^{e} the tetragonal phase was reported in [251], another relevant phase observed in nanoparticles ([248],[252]) is the cubic Pd$_{17}$Se$_{15}$ [253]; ^{f} tetragonal: [254], orthorhombic: [255]; ^{g} observed in nanorods [245], another important phase is the orthorhombic one ([256], JCDPS # 00-024-1041); ^{h} [257].

3.2 Metal-CdSe nanopyramid hybrid structures - deposition and ion exchange

One important aspect in the seeded-growth formation of hybrid nanoparticles is the morphology of the seed. The type of ion terminating a crystal facet (for example Cd^{2+} or chalcogenide) and the number of dangling bonds lead to inequalities of surface positions and preferential deposition of metals or second materials in general [258, 98].

CdSe nanopyramids with wurtzite structure are an interesting model system to study metal deposition related effects due to their specific surface structure. This structure provides several defined crystallographic sites with high reactivity as emphasised in Figure 3.4. Additionally, facets of high roughness and better accessibility due to a mixed coordination of long-chained ligands and chloride are present (see Chapter 2). These facets may also appear in smaller dimensions at defect sites of rods and at their tips [230]. In a hexagonal dipyramid with large $\{10\bar{1}1\}$-type facets the majority of the crystal may be terminated by one type of ion. With strong X-type ligands a Cd-rich surface is favoured but etching of the topmost layer would result in a Se-rich configuration, which may alter the deposition behaviour of metals significantly.

3.2 Metal-CdSe nanopyramid hybrid structures - deposition and ion exchange

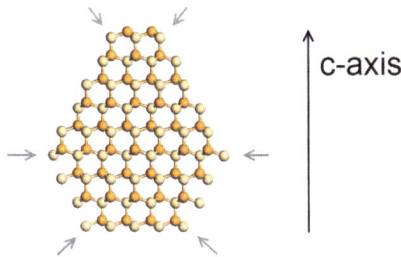

Figure 3.4: Bond scheme of an ideal hexagonal CdSe dipyramid, the arrows point to reactive sites with the highest number of dangling bonds, where metal deposition is expected. Beige atoms represent Cd, orange atoms are Se.

CdSe and Au is a model system for deposition studies of metals onto semiconductor nancrystals. With CdSe nanopyramids, the formation of an unstable shell-like layer of Au on the nanoparticles was observed by the author and co-workers [105]. The only other example of a shell-like metal structure on hexagonal CdSe nanoparticles contains partially pyramidal seeds and Pt [259]. In that case, the metal was deposited in water after a ligand exchange to hydrophilic stabilisers on CdSe. No details about the formation mechanism were provided but an enhanced catalytic activity compared to commercial Pt catalysts in direct methanol fuel cell reactions was claimed.

For a further application of the hybrid material a deeper understanding and control over the formation or prevention of the Au shell are necessary. Thus, studies on its deliberate preparation, configuration and instability against the electron beam of the TEM were carried out. The latter aspect touches on a general difficulty in hybrid and metal nanoparticle analysis with electron sources, which can lead to misinterpretations of particle morphologies.

To derive a more general insight into metal deposition on hexagonal CdSe nanopyramids and eventually obtain a stable oligomeric hybrid structure, reactions with silver, palladium and platinum precursors were examined additionally.

CHAPTER 3. METAL-SEMICONDUCTOR HYBRID NANOPARTICLES

3.2.1 Au-CdSe nanopyramid structures

Preliminary work on the deposition of Au onto CdSe nanopyramids in organic solution revealed a peculiar behaviour different from the one observed with rod-shaped or spherical seeds [17]. Incubated with an Au precursor solution containing gold(III)-chloride ($AuCl_3$), n-dodecyltrimethylammonium bromide (DTAB) and ligands such as the long-chain amines hexadecylamine (HDA) or oleylamine (OAm) in toluene, a shell like Au-structure formed on the surface of CdSe nanopyramids [17, 105]. The presence of this layer could be observed indirectly in form of Au domains growing on the surface of the nanopyramids when examined with special care, low beam currents and quickness in a transmission electron microscope (TEM). Such changes under the influence of an electron beam were not registered when a reducing agent, tetra-n-butylammonium borohydride (TBAB), was added to the solution after Au deposition. This points to unreduced Au species in the shell, especially because dot-shaped Au domains formed with the reducing agent. Heating of the samples during the reaction did not affect the shell and its instability against the electron beam. An overview of these connections is shown in Scheme 3.1.

The size of the reduced Au-dots was varied between 1.4 and 3.9 nm by changing the Au to CdSe ratio and the ligand. The smallest Au dots were homogeneously distributed on reactive CdSe sites as demonstrated by (scanning) transmission electron microscopy (Figure 3.5). They were obtained with dodecanethiol (DDT) as ligand for Au. A strictly epitaxial deposition could not be confirmed but a direct nucleation of Au onto the CdSe lattice with strain relieving edge dislocations on the side of Au is visible in the inset in Figure 3.5 c.

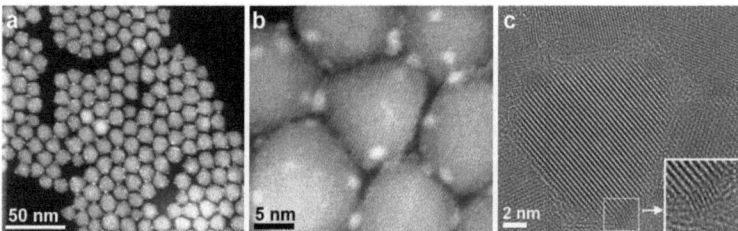

Figure 3.5: CdSe nanopyramids decorated with cluster sized Au domains. (a, b) Electron micrographs obtained in the scanning transmission mode (STEM) and (c) high resolution TEM micrograph with enlarged view of an interface region between CdSe and Au (visible lattice planes: CdSe (0002), Au (200)).

3.2 Metal-CdSe nanopyramid hybrid structures - deposition and ion exchange

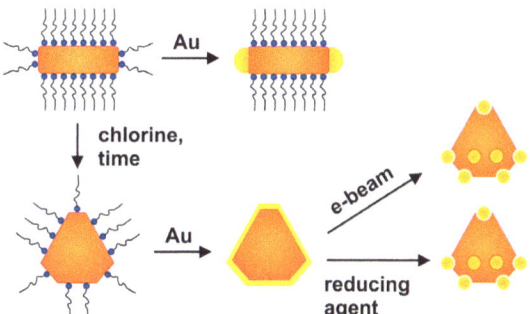

Scheme 3.1: Scheme of Au deposition onto CdSe nanorods and pyramids with amine ligands. The distinct morphology and mixed ligand sphere of the latter make the surface more accessible for Au-deposition. A shell-like structure is formed, which is unstable against electron beam irradiation and addition of a reducing agent [105] - Reproduced by permission of The Royal Society of Chemistry.

In the current work it could be shown that Au precursor with dodecanethiol also produced spherical deposits without the addition of a reducing agent. In an experiment with a stepwise increase of the Au content, ripening of the metal domains from several distributed to few large ones occurred. Remarkably, with higher amounts of Au slight growth of the metal domains appeared to happen during TEM inspection. The morphological evolution in solution is similar to the one found in experiments with amines and reminiscent of processes observed with spherical and dot shaped semiconductor seeds [17, 168, 228]. It is attributed to an electrochemical Ostwald ripening, where smaller domains are oxidised and the electrons are transferred across the semiconductor to a larger domain where gold precursor is again reduced.

In Figure 3.6 TEM micrographs and absorbance spectra with increasing Au to CdSe nanopyramid ratios are shown (CdSe nanopyramids were prepared with 1,2-dichloroethane, DCE/Cd 0.7, see Chapter 2). The most reliable and feasible method to determine a value for the latter was found to be the ratio between the molar amount of metal and a *quasi* molar amount of CdSe nanoparticles. This *quasi* amount is the product of volume and optical density of the employed CdSe dispersion. An estimation of the corresponding CdSe content and real molar ratios was carried out by following the method of Leatherdale *et al.* [260], as described in detail in the experimental section (Chapter 5).

CHAPTER 3. METAL-SEMICONDUCTOR HYBRID NANOPARTICLES

The maximum ratio of Au/CdSe in Figure 3.6, 7.30, thus corresponds to an estimated molar excess of 8400 (7.9×10^{-6} mol of Au and 9.4×10^{-10} mol CdSe nanopyramids)[9]. With higher loads of Au, the absorption features become smoother. While, similar to samples with Au shell [17], no plasmonic resonance of Au was observed, a tail towards longer wavelengths arises and the absorbance at lower wavelengths is slightly increased. The latter may be attributed to interband transitions in the metal [261], while the changes of the absorption maxima belonging to the semiconductor are explained by electronic interactions across the interface as referred in Section 3.1.1. From the observed deposition and ripening behaviour the question arose why such differences occur between incubation with Au precursor solutions containing amines and those containing dodecanethiol. The impact of thermal energy and photogenerated electrons on the hybrid morphology were tested. In reactions with Au-amine or pure stock solutions, a shell formed even under applied heat [17, 105] or UV-irradiation[10] during the deposition. Based on these findings, the responsible factor was presumed to relate to the Au precursor and the role of the reagents was examined more closely.

---

[9] The complexity of the pyramidal geometry requires approximations in the calculation of volumes needed for the determination of concentrations. Additionally, the two-dimensional projection in the TEM micrographs limits the accuracy of determined dimensions. For these reasons larger deviations may occur between different batches of nanoparticles. Based on the *quasi* molar amount the formed metal domain sizes were in good agreement between batches.

[10] In a cuvette, CdSe nanoparticles were irradiated with UV-light of 366 nm before and while Au(III)-stock solution was added under stirring. An Au shell deposited onto CdSe nanopyramids.

3.2 Metal-CdSe nanopyramid hybrid structures - deposition and ion exchange

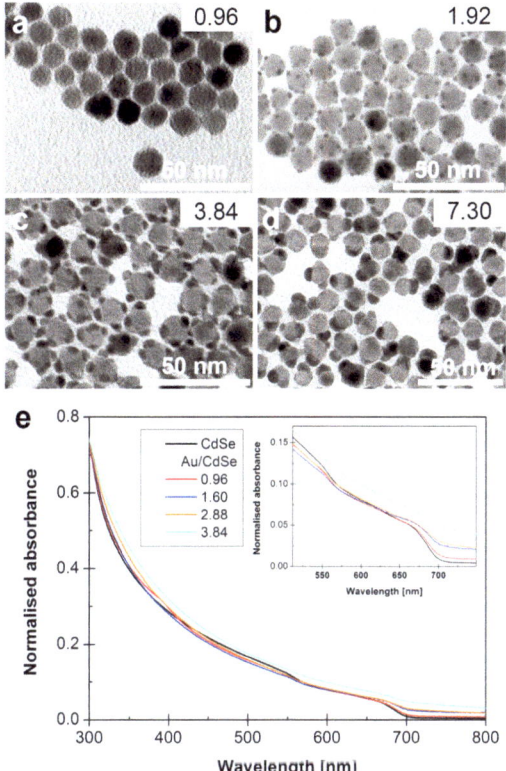

Figure 3.6: (a - d) Au domain growth and (e) absorption spectra with increasing Au/CdSe ratio. The CdSe nanopyramids were prepared with 1,2-dichloroethane, the gold solution contained dodecanethiol as ligand.

CHAPTER 3. METAL-SEMICONDUCTOR HYBRID NANOPARTICLES

3.2.1.1 Influence of the precursor oxidation state on the hybrid morphology

The choice of Au precursor is of prime importance in order to obtain defined semiconductor-gold structures. As Au needs to be reduced to its elemental state during hybrid formation, the oxidation state of its precursor (Au(III) or Au(I)) can make a decisive impact on the geometry and stability of the later formed hybrid nanoparticles. In lead sulphide, for example, Huang *et al.* demonstrated that a Au(I)-precursor was reduced to form metallic domains on the surface of the seed material, while Au(III)-precursor rather destructed the nanoparticle [172]. The oxidation state is not only determined by the Au source, often $AuCl_3$ or auric acid ($HAuCl_4$), but also by ligands which may act as mild reducing agents. Au precursors commonly employed for the deposition onto CdSe nanoparticles are combinations of Au(III) with alkylammonium bromides and amines such as dodecylamine [10, 161] or Au(I) components prepared with thiols [168]. In the stock solutions applied here, $AuCl_3$[11] was complexed with dodecyltrimethylammonium bromide in toluene under excess of the latter (1:1.5). Ammonium halides and Au(III) form complexes of the type $DTA^+[AuX_4]^-$ [262, 263]. A mixed coordination by Cl^- and Br^- can be assumed, owing to the higher likeness of soft Au and Br ions [264]. When the orange Au(III) stock solution was shaken for 5 minutes with dodecanethiol or oleylamine in a 23-fold excess, it turned colourless with the thiol while it remained pale yellow with oleylamine (Figure 3.7) [17]. After 24 hours the lack of colour indicated that both solutions contained Au(I) at that point [265], while Au in the stock solution remained unreduced. This documents that the amine reduces Au(III) much slower and that there is a mixture of Au(I) and Au(III) after 5 minutes (see also references [266, 267]), which is a typical period between preparation and injection of the solution in methods based on Mokari's recipe [10].

The oxidation state of the precursor employed determines the number of electrons necessary for the formation of Au(0), thus differences in deposition behaviour may be expected. Indeed, substantial dissimilarities in the deposition of Au onto CdSe nanopyramids arose upon incubation with solutions containing Au(III) or Au(I). For these examinations nanopyramids synthesised with 1-chlorooctadecane served as seeds. They were prepared after the recipe described in Chapter 2 with injection of the additive in a ratio of 0.7 to Cd just before the injection of TOPSe and a growth period of 24 h. With such a long reaction time the residue of precursors could be reduced, noticeable for instance by the absence of a diffraction peak of Cd-phosphonates at $2\theta = 20°$ [268]. This small peak was

---

[11] The advantage of $AuCl_3$ compared to $HAuCl_4$ is that no acid, which may etch CdSe, is produced during complexation.

3.2 Metal-CdSe nanopyramid hybrid structures - deposition and ion exchange

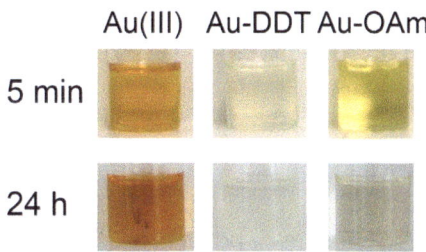

Figure 3.7: Au precursor solutions prepared without additional ligand (Au(III)), with dodecanethiol (DDT) and with oleylamine (OAm), 5 minutes and 24 hours after mixing.

often visible with pyramids synthesised with DCE and took several steps of purification to be removed. A minimum amount of residues was desired for the analytical experiments shown below. A difference between the obtained nanoparticles with dimensions around 10.5×12.0 nm and those prepared with DCE was their hexagonal monopyramidal instead of dipyramidal structure (the smaller frustum was hardly or not developed). Compared to DCE-prepared nanoparticles a lower number of crystal vertices was thus available but this did not alter the general deposition behaviour.

Same as above, a CdSe dispersion with defined optical density was prepared and mixed with a gold-precursor solution. No difference in the reaction outcome was observed between ambient and inert gas conditions, so that ambient conditions were continuously applied. The total concentration of nanoparticles and Au precursor played a role rather in terms of colloidal stability than for the morphology of the products in the range of examined volumes. Shell type hybrid nanoparticles, for example, were more easily re-dispersed in smaller volumes of solvent. Sampling at different stages required larger volumes. Because of this, two recipes were developed, one diluted and the other one more concentrated. In the concentrated version, undiluted volumes of the Au precursor solution were injected to CdSe nanopyramid dispersions of 0.5 mL with an optical density of 1.60. For the diluted recipe, Au precursor solutions in 1 or 2 mL of toluene were injected into CdSe receivers of 2 or 4 mL with an optical density of 0.27. Regarding the morphology of the products, the total concentration was found to be far less important than the Au to CdSe ratio, which ranged between 1.4 and 3.9.

CHAPTER 3. METAL-SEMICONDUCTOR HYBRID NANOPARTICLES

In Figure 3.8 high resolution micrographs of hybrid nanoparticles resulting from the reaction between CdSe and Au(III)- and Au(I)-precursor solution are shown. Nanoparticles incubated with Au(III)-solution precipitated and agglomerated but were re-dispersed by addition of an excess of ligands such as oleylamine or dodecanethiol (at least 1:23 with oleylamine and 1:18 with dodecanethiol)[12]. Added ligands allowed for cleaning processes that were essential for high resolution electron microscopy. The growth of solid Au domains from a shell-like layer during TEM inspection is shown in Figure 3.8 a and b. The polycrystallinity of the formed Au-domains becomes evident from the lattice fringes visible in the micrograph in Figure 3.8 d. At cryogenic conditions (77 K) the movement of Au atoms was reduced so that the shell structure could be recorded in the scanning (inset in Figure 3.8 a) and normal TEM mode Figure 3.8 c. As expected from prior work, shell formation was also observed with freshly prepared oleylamine solutions. Remarkably, the migration of Au was influenced by the ligands, since hardly any movement was observed in samples prepared with only Au(III) compared to those with extra ligands (not shown). The latter examinations were carried out at lower magnification and acceleration voltage (100 kV instead of 200 kV)[13].

Standard reactions with Au(I) were conducted with Au-DDT solutions and yielded dot-shaped Au deposits on crystallographically exposed sites on CdSe (Figure 3.8 e). Through the deposition it became clear that the employed CdSe seeds prepared also contained a fraction of tetragonal pyramids. This did not affect studies on the deposition mechanism. Nevertheless, a lack of homogeneity limits the applicability of the structures for studies of properties, for example concerning electrical transport. At the higher Au/CdSe ratio of 2.80, growth of the Au domains was observed during TEM inspection. A discussion of this aspect will follow later in the text.

In conclusion, two different morphologies of hybrid nanoparticles were accessible with the same ingredients: CdSe nanopyramids, Au(III)-precursor solution and dodecanethiol. By simply changing the sequence of mixing them the precursor oxidation state can be controlled which determines the mode of deposition.

---

[12] Oleylamine only loosely stabilises the nanoparticles. After precipitation for cleaning they became unstable again. With dodecanethiol several cycles of precipitation and re-dispersion were possible.

[13] There is a dependence of the Au domain growth on the beam current as shown in Appendix A.

3.2 Metal-CdSe nanopyramid hybrid structures - deposition and ion exchange

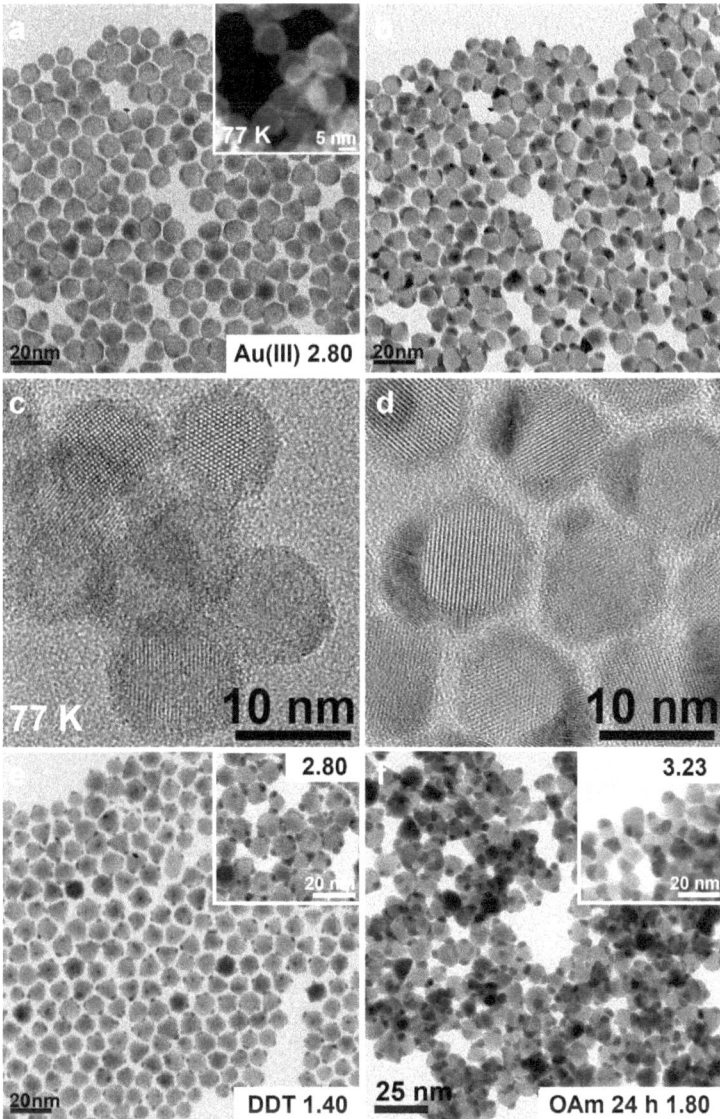

Figure 3.8: (a-d) HR-TEM micrographs of Au-CdSe shell samples prepared with pure Au(III)-solution. The inset in (a) shows an STEM image of a shell sample at cryogenic conditions. (e) Samples prepared with Au-DDT (Au(I)) at different Au/CdSe ratios. (f) Samples incubated with Au-OAm precursor aged for 24 hours after mixing (Au(I)) at different Au/CdSe ratios.

59

CHAPTER 3. METAL-SEMICONDUCTOR HYBRID NANOPARTICLES

Control experiments were undertaken to prove that the oxidation state of the precursor is the driving force behind the difference in morphology. First, Au-OAm precursor solutions were aged for 24 hours before the colourless liquid was added to the CdSe receiver at different Au/CdSe ratios. The resulting hybrid nanoparticles with spherical Au-domains are shown in Figure 3.8 e.

To further rule out a stronger seed passivation by excess dodecanethiol, CdSe nanopyramids were treated with the ligand under conditions similar to Au-deposition but without Au. The nanoparticles were stirred with the same concentration of ligand as usually added with the Au precursor (diluted method, mol DDT:mol CdSe nanopyramids = 65958:1). After five minutes they underwent three cycles of precipitation and re-dispersion with toluene/ethanol to remove unbound organics. Small shifts and few additional signals compared to untreated samples appeared in ATR-IR-spectra (Figure 3.9). Phosphorous ligands were still present as identified by characteristic $\nu(P=O)$, $\nu(P-O)$ and $\nu(P-O-(H))$ bands [269]; the characteristic $\nu(C-S)$ at $655\,\text{cm}^{-1}$ of dodecanethiol that should be visible in nanoparticle bound thiols did not appear [69, 270]. A hindered or incomplete ligand exchange is not unusual with thiols when no base is supplied to deprotonate the headgroup [69]. Afterwards, the nanoparticles were incubated with Au(III)-precursor, which resulted in shell formation. These findings indeed confirm that a passivation of the seed surface by dodecanethiol is not the major reason for the formation of dot-shaped domains.

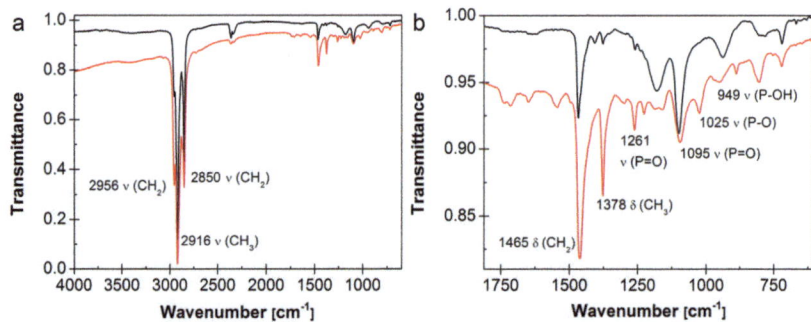

Figure 3.9: IR-spectra of CdSe nanopyramids prepared with 1-chlorooctadecane before (upper curve) and after treatment with dodecanethiol (lower curve). In (b) the lower wavenumber regime of the spectra in (a) is enlarged. After the treatment, $\nu(P\text{-}O)$ bands are present and in some cases even more pronounced. Apart from the CH$_3$ band at 1378 cm^{-1} no signals attributable to dodecanethiol appear.

3.2 Metal-CdSe nanopyramid hybrid structures - deposition and ion exchange

Having established that it is the precursor oxidation state that governs the deposition, different possibilities arise for the configuration of the shell. The first one is that it contains oxidised Au species due to incomplete reduction of the higher valent Au(III) precursor. This would fit with the demonstrated behaviour under electron beam irradiation and results of Mokari et al. who mentioned a TEM induced formation of Au domains on aggregated Au-CdSe nanodumbbells prepared with an Au(III) solution if dodecylamine was not added [173]. Another example in the same direction is the growth of Au domains on the surface of CdS nanorods from adsorbed Au-oleylamine complexes also upon interaction with the electron beam [271]. Ripening and nucleation phenomena in the TEM are furthermore observed with ligand bound Au(I) species in crystallised complexes with oleylamine [272]. Oxidised Au on the surface of Au(0) nanoparticles is also intensively discussed in regard to the misinterpretation of TEM data [273, 274]. The shell might thus fully consist of an oxidised Au compound or contain an outer layer of unreduced species. Such a layer would also explain the electron beam induced growth of dot-shaped deposits prepared with Au-DDT precursor at higher Au/CdSe ratios.

Another possibility is that the shell is fully reduced to Au(0) but metastable simply due to its small dimension, which ranged below 1 nm where visible, or due to an amorphous structure. Also nanoscaled Au is capable of rearranging or even melting upon interaction with an electron beam [275]. Anyhow, with this option the observed effect of the reducing agent added in solution would be more difficult to explain.

A first clue to this might be derived from X-ray diffraction (XRD) patterns. Those of samples with Au dots were the same as of the starting CdSe nanopyramids, indicating that the crystalline Au domains were too small to be recorded. Samples with Au shell showed a small difference towards the pristine material (Figure 3.10). There is a new broad signal between $2\theta = 29°$ and $34°$, whose identification bears some ambiguity since it could be an overlap of several peaks or broadened due to nano size effects. Characteristic Au diffraction peaks would appear at $2\theta = 38°$ and $44°$ (JCPDS # 00-004-0784), which is not the case. Thus, the shell is not crystalline, too thin or not fully composed of Au. Furthermore, patterns of Cd, Se, their oxides or mixed compounds of Cd and Au do not comply. Precursor related compounds similar to $AuCl_3$ would exhibit recognisable peaks at $2\theta = 20°$ (JCPDS # 01-073-0370). The diffraction pattern of monoclinic α-AuSe (mixed Au(I) and Au(III) [250]) has its maximum peaks in the region of the additional peak and is a possible candidate to form on the surface of the CdSe nanopyramids. Nevertheless, the mentioned uncertainties remain.

CHAPTER 3. METAL-SEMICONDUCTOR HYBRID NANOPARTICLES

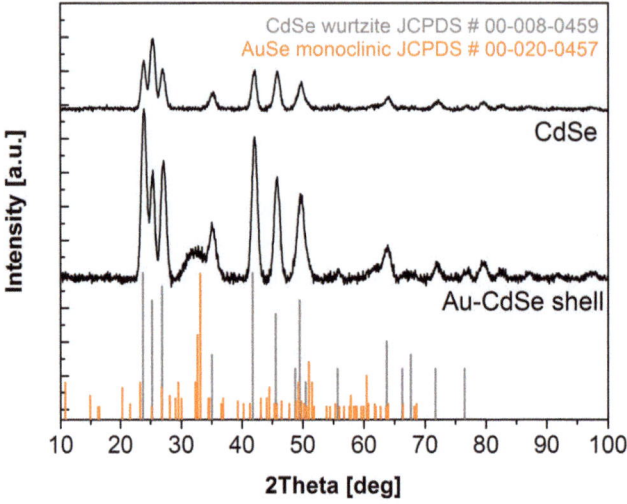

Figure 3.10: The X-ray diffraction pattern of Au-CdSe shell nanoparticles contains the original CdSe wurtzite signals and a new component with broad peaks similar to monoclinic bulk α-AuSe.

For applications such as catalysis it is important to know the oxidation state and composition of the system, especially of the metal containing component, since it determines the type of reaction that can be catalysed. Positive oxidation states may act as catalytic moieties in organic coupling reactions and improve the conversion compared to elemental Au domains, as demonstrated for Au(I) species on the surface of Au-PbS nanoparticles [175]. In electrical transport experiments, on the contrary, positive oxidation states can lead to charging effects which influence the characteristics significantly.

For these reasons, further investigations were carried out to reveal the constitution of the Au shell and whether it was bound to the nanoparticle surface or rather loosely adsorbed to it.

3.2.1.2 Compositional analysis by EDX and XPS

Differences in the composition and oxidation state of Au-CdSe samples with dots or shell were examined by EDX and XPS. CdSe nanopyramids prepared with 1-chlorooctadecane were employed and the incubation was carried out by direct mixing of Au and CdSe solutions with the concentrated method.

3.2 Metal-CdSe nanopyramid hybrid structures - deposition and ion exchange

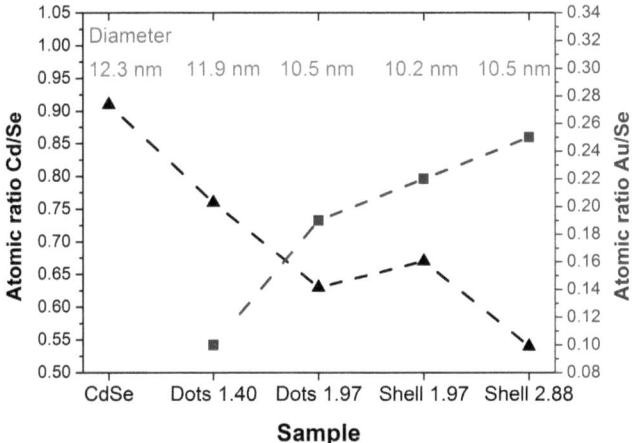

Figure 3.11: Atomic ratios of Cd to Se (triangles) and Au to Se (squares) calculated from EDX data. The diameter of the samples as determined from TEM is written above the data points. The standard deviations of these values are between 1.1 and 1.3 nm.

EDX measurements were conducted with a row of experiments to determine the relative Au content and changes in the Cd to Se ratio. Hybrid nanoparticles prepared with Au(I)-precursor at Au/CdSe ratios of 1.40 and 1.97 as well as samples incubated with Au(III)-solution at Au/CdSe ratios of 1.97 and 2.88 were examined. The Au dot size with the smaller ratio was 3.1 ± 0.9 nm, the larger ratio lead to dots with 4.0 ± 0.7 nm. The latter exhibited slight Au-migration during TEM inspection. Both shell samples behaved as shown in Figure 3.8. The changes in the shell structure prevented statistics of its thickness for a correlation with the employed Au/CdSe ratio. General trends in the data visualised in Figure 3.11 are an increase of the Au to Se ratio[14] with higher amounts of Au available and with a shell compared to dots. Concomitantly, the Cd to Se ratio decreased. The deposi-

---

[14] In contrast to TXRF and ICP-OES analyses (own and ref. [125]), there is a small excess of Se found in EDX date of pure CdSe nanopyramids. Additionally, there are variations in the Cd to Se ratio of pure CdSe in Figure 3.11 and Tables 3.3 and 3.4. The latter result from intrinsic differences in the detection system and analytical software of the microscope-EDX systems used for recording the data. The data for Au was obtained on a Philips CM 300 UT with an EDAX DX-4 system, operated at 200 kV, while the other data was measured with a JEOL JEM 2200FS (UHR) microscope with a JEOL JED-2300 EDX detector. Variations between values obtained with differing analytical methods may arise from their specific sensitivities for certain elements and alterations in the sample preparation. Even though this hampers the determination of the true Cd to Se ratio, the results were consistent within one method, so that trends during reactions with metals could be monitored by EDX.

tion of Au etched the CdSe seeds, irrespective of the employed precursor, as the diameter of the nanoparticles shrank. Micrographs of the samples and changes during TEM inspection are shown in Appendix A. Au preferentially withdrew to the apex during TEM inspection and covered parts of the nanoparticles from sight. This prevented a reliable determination of the c-axis length in shell samples so that the diameter is the sole dimension given in Figure 3.11. Such etching processes were observed in other reports and traced back to a dissolution of nanoparticle surface atoms after reduction of Au by chalcogenides [223, 229] or by solvation through ligands [10]. With distances between lattice planes perpendicular to the c-axis of 0.37 nm ($1\bar{1}00$) and 0.22 nm ($11\bar{2}0$) [60], the differences between the diameters of CdSe nanopyramids imply that more than one atomic layer is lost after Au deposition at Au/CdSe ratios higher than 1.40. In combination with the circumstance that the Au(III)-absorption band of the stock solution vanishes within seconds after the addition of CdSe nanopyramids [17], these findings point to a reduction reaction between Se(-II) and Au(III) with concomitant dissolution of Cd(II).

Even though there seems to be a reduction of Au precursor by the nanoparticles, the shell must not necessarily be reduced throughout. Nevertheless, an indication of Au(0)-deposits was derived from cyclic voltammetry in organic solution. In ongoing studies carried out by Leonor de la Cueva at the Universidad Autónoma de Madrid involving the hybrid nanoparticles, no reduction of oxidised Au species was observed. This means that little amounts may be present but a shell completely consisting of oxidised Au is not likely.

To further clarify this point, the oxidation states of the Au components in CdSe-Au dot and shell samples were investigated by synchrotron XPS. The spectra were measured by Roberto Otero of the Universidad Autónoma de Madrid and IMDEA Nanoscience Madrid and co-workers from Hamburg and Madrid and fitted by the prior. Samples of CdSe and the two extreme cases of CdSe-Au shell and CdSe-Au dots nanoparticles prepared at Au/CdSe ratios of 2.88 and 1.40 were examined.

Survey spectra of the three, obtained at a photon energy of 720 eV with an energy pass of 50 eV, are shown in Figure 3.12. There are no phosphorous or chlorine signals in the survey data of samples with Au. This implies that the original ligands of CdSe were largely removed or exchanged for Au compounds in the case of the shell. A lack of signals in the regions of chlorine and bromine binding energies (199 eV and 69 eV) indicates that halides are not incorporated into the Au shell in significant amounts even though available in the precursor solution ($AuCl_3$, dodecyltrimethylammonium bromide), which excludes the possibility of AuCl or AuBr-related compounds on the surface. Anyhow, a presence of

3.2 Metal-CdSe nanopyramid hybrid structures - deposition and ion exchange

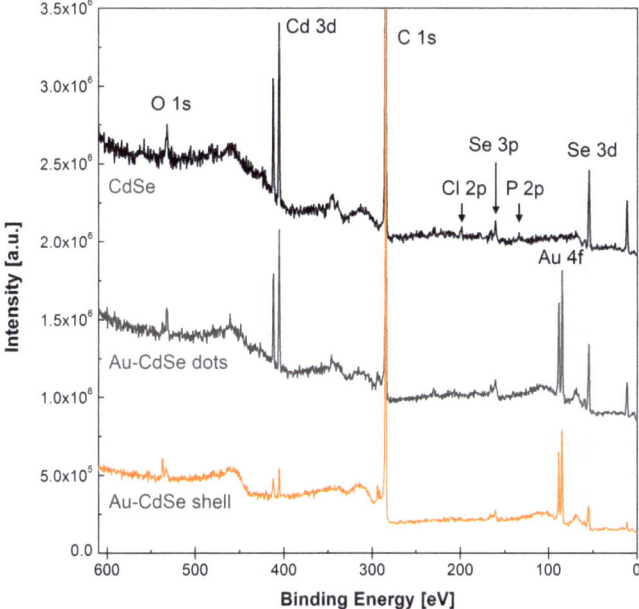

Figure 3.12: XPS survey spectra of CdSe nanopyramids and Au-CdSe hybrid nanoparticles with dot shaped domains and shell. The photon energy was 720 eV and the energy pass was set to 50 eV. The relevant Au and Se peaks are marked.

the halides cannot be excluded completely, as the sensitivity factors of the relevant orbitals are low compared to Au and they might thus be hard to detect. Oxygen signals usually arise from environmental oxygen adsorbed during sample transfer.

In high resolution scans the Cd signals were not shifted significantly. The high resolution signals relevant for the analysis of the interaction between nanoparticle surface and Au, Se 3d and Au 4f, are compiled in Figure 3.13. In both Au-CdSe samples the Se 3d signal can be deconvoluted into at least two components. The Se peaks centring at 59 eV in the Au-CdSe samples can be related to a partial oxidation during transport or transfer of the sample to the measuring set-up, since the survey spectra contain broader oxygen signals at 532 eV. The oxidised species rather stem from Se residues or Se(0) species remaining on the substrate than nanoparticle bound ones. The reason for this assumption is that the fitted Se components are very similar to those in another Au-CdSe sample, where no

CHAPTER 3. METAL-SEMICONDUCTOR HYBRID NANOPARTICLES

oxidation had taken place but the corresponding data-set is not presented as it remained incomplete for technical reasons (the Se signal is shown in Appendix A).

The bulk signals, recognisable from their smaller full width at half maximum (FWHM, < 1 eV), are slightly shifted towards lower binding energies. From 54.1 eV in the original CdSe nanoparticles the Se 3d 5/2 peak moved to 53.8 eV in samples with Au. Such a shift is unexpected since it means that there is a higher electron density on the atom. Anyhow the difference is very small and a similar shift is visible in XPS data of Au-CdSe nanodumbbells in reference [13], albeit with lower overall resolution, where it is denoted as insignificant.

In the present data the second Se 3d 5/2 contribution, appearing in presence of Au, is located towards higher binding energies at 54.3 eV, where electron density is removed from the atom. This contribution lies too low to be assigned to Se(0) according to available reference data that ranges between 54.7 and 57.6 eV [276]. If the areas of the two contributions with lower binding energy are counted as 100%, the second contribution is much stronger in the case of the shell sample than with Au dots, with 68% compared to 35%. This suggests a correlation to the degree of coverage of the CdSe surface by Au atoms and supports the assumption that there is a direct interaction or bond between surface Se atoms and Au. It should be noted that Au may shield electrons from the core due to its high absorption cross section, which could influence the relative intensities of the components. The fits of the second Se component in both dot and shell samples are broader than the reference Au signal (Se 3d 5/2 shell: 1.2 eV; Se 3d 5/2 shell: 1.1 eV; Au 4f: 0.8 eV), which hints at the existence of further Se components with slightly different chemical environments. These may be related with the distance towards the Au domains.

The major Au component in both hybrid nanoparticle samples was assigned to Au(0) with a binding energy of 84.0 eV for the Au 4f 7/2 photoelectron line. Fitting of a second component revealed differences between the samples. With Au dots, the fit is centred at 84.5 eV with a FWHM of 1.3 eV and contributes 12% of the total area of the Au 4f 7/2 photoelectron line. The peak position is in agreement with those reported for Au(I)-thiolate complexes and surface bound Au(I)-thiolate layers on Au nanoparticles [277, 278]. A contribution from Au-Se interactions may be part of this comparatively broad signal. In the shell sample the maximum of the second component is positioned at 85.0 eV and very broad, with a FWHM of 1.9 eV. It contains 14% of the area of the Au 4f 7/2 peak. Judging from the large width of the fit there are more components underneath but the low intensity of the signal at this binding energy prevents reliable splitting into further fits. The different chemical environments for the Au(I) species may be surface related but

3.2 Metal-CdSe nanopyramid hybrid structures - deposition and ion exchange

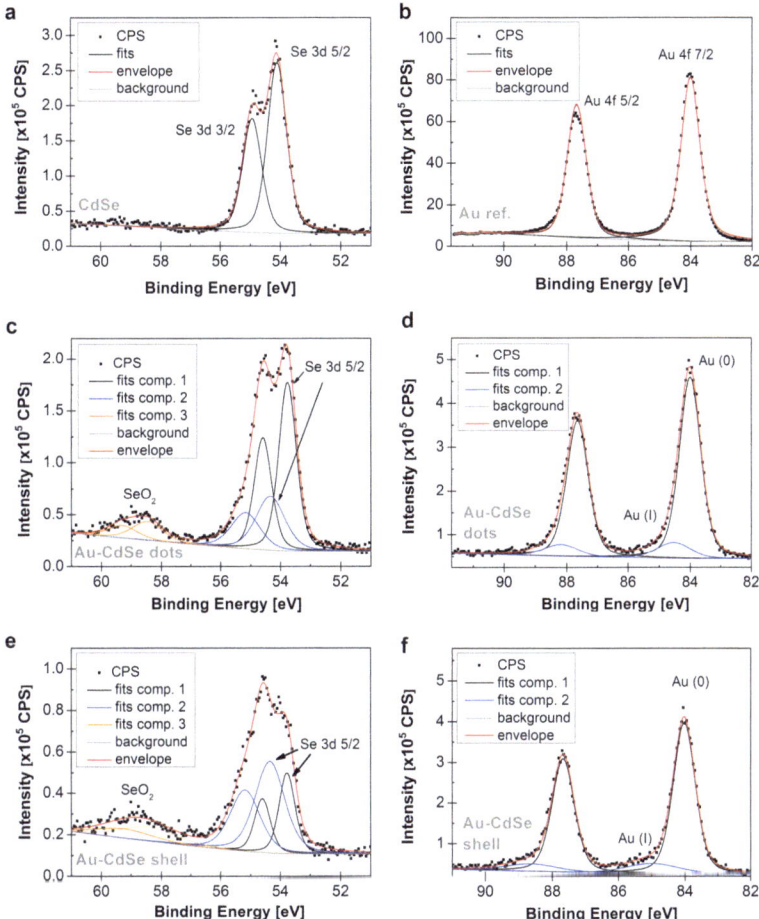

Figure 3.13: (a), (c), (e) Se 3d XPS signals and fits of the original CdSe nanopyramids and Au-CdSe samples with dots (Au/CdSe 1.40) and shell (Au/CdSe 2.88). (b) Au 4f reference signal and (c, f) Au 4f regions of the Au-CdSe samples with fits. The photon energy was 720 eV and the energy pass was set to 20 eV. The difference in the Au(I) components between the hybrid structures and the appearance of a new component of Se indicate a stronger interaction between Au and Se in the shell sample.

also be caused by an interaction with Se. A presence of Au(III) species is unlikely since corresponding signals as with [NR$_4$][AuX$_4$] are expected between 67.7 eV and 87.6 eV [279]. With this, the presumed Au^{I}Au^{III}Se phase found in the XRD pattern of Au-CdSe shell samples seems to be formed as a side product in the reaction that is not present in the XPS sample.

Further distinction between the Au species and their location would require the identification of thiolate bound Au(I) surface atoms through an analysis of the S 2p peak. The signal for sulphur is positioned at the same binding energies as the Se 3p signal, which is composed of at least two components itself. Together with the low intensity of the signal at a binding energy of 155 eV, a serious deconvolution is not feasible under the given circumstances.

In summary, the XPS data confirms a direct interface between Au and CdSe via interaction with Se atoms and reveals the mostly elemental state of Au, with below 15% of oxidised Au(I) species in both morphologies. The position of the Au(I) peak in nanoparticles with Au dots fits with surface bound Au(I)-thiolates. In samples with an Au shell, the higher interface area between Au and Se has a stronger impact on Se atoms in the nanoparticle, which is interpreted as a higher number of direct bonds. The width and position of a fitted Au(I) compound suggest the presence of several small contributions from Au components with slightly changed chemical environments that may be located on the surface and/or at the interface towards CdSe.

3.2.1.3 Reasons for the different deposition behaviour

The above results show that Au(I) can easily deposit onto reactive sites of CdSe nanopyramids under withdrawal of electron density from Se and dissolution of Cd, which is in agreement with mechanisms suggested for other metal-Cd chalcogenide systems [11, 100, 160, 229]. Further growth to spherical domains occurs quickly and can be understood by a self-catalytic reduction on the formed clusters aided by aurophilic attractions between Au(I)-precursors [280, 281]. Furthermore, CdSe nanopyramids are able to reduce Au(III) to Au(I) and Au(0) under the formation of Au-Se bonds resulting in a core-shell structure with the majority of Au atoms in the elemental oxidation state. In regard to the difference in the deposition of Au(I) and Au(III), there seems to be a relation to the rate of the reduction in analogy to reference [225]. De Paiva *et al.* carried out calculations concerning the likelihood of the formation of an Au layer on different CdSe surfaces [282]. According to their results, a full Au layer is energetically favourable only under Au-rich conditions

3.2 Metal-CdSe nanopyramid hybrid structures - deposition and ion exchange

and when the surface contains Cd-vacancies or is rich in Se. This fits well with the observed loss of Cd and explains the unique ability of pyramidal CdSe nanoparticles to support a shell of Au. In contrast to rod-shaped nanoparticles, pyramids may be easily etched to become Se-rich on the whole surface due to their sloped $(10\bar{1}\bar{1})$ facets (see Chapter 2). The calculations also stated that Au influences the crystal structure of CdSe mostly in the first and second layer. Such a change in the environment of Se atoms would contribute to the observed differences in the XPS signals. The layer formation with Au(III) can thus be explained by its lower reduction potential (see Table 3.1) and weaker interactions between precursor complexes which enables a removal of Cd under Au-rich conditions and filling of Cd vacancies by Au.

At large ratios of Au(I) to CdSe, there is a maximum Au dot size between 3 and 4 nm above which ripening and the formation of unstable layers become more feasible, as observed by growth processes under the electron beam. An excess driven deposition on the remaining CdSe surface is equally likely as a higher content of ligand bound Au(I) on the surface or a combination of the two. A deconvolution of the XPS signal of S would help to differentiate between Au-thiolate and other Au species but was not feasible. Nevertheless, a surface coverage of the Au domains by polymeric Au-thiolates could explain reduction and migration processes under the TEM [273].

To better understand the migration of Au atoms with the knowledge that the shell structure contains mostly Au(0) and to possibly affect the formation of networks of Au-connected CdSe nanopyramids inspired by the diffusion observed in the TEM and reference [169], Au-CdSe shell monolayers were subjected to thermal treatment.

3.2.1.4 Annealing of Au-CdSe shell monolayers

An attempt was made to exploit the instability of the Au shell to form networks between hybrid nanoparticles by *ex situ* annealing *in vacuo*. CdSe nanopyramids prepared with DCE were first incubated with Au at a Au/CdSe ratio of 1.92 and re-dispersed with the help of dodecanethiol. The resulting hybrid nanoparticles were purified and assembled on diethyleneglycol by the Langmuir-Blodgett technique. Afterwards they were deposited onto a silicon substrate with a 300 nm silicon oxide layer. The method was based on the preparation of CdSe nanopyramid monolayers by Cai [132]. Afterwards, the substrates in form of silicon wafers and TEM grids were heated in a tunnel furnace to temperatures from 70 to 200 °C. Each sample was heated to one specific temperature. In TEM micrographs in Figure 3.14 no effect is visible after annealing to 100 °C, the typical migration of

CHAPTER 3. METAL-SEMICONDUCTOR HYBRID NANOPARTICLES

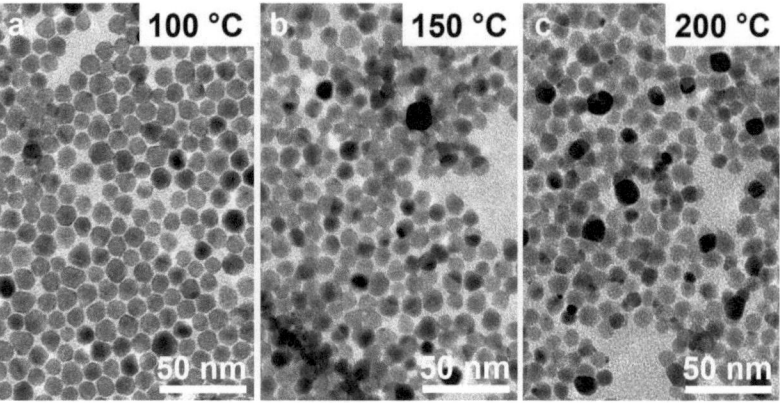

Figure 3.14: TEM micrographs of Au-CdSe shell nanoparticle films after annealing at different temperatures. Au forms large particles through heating.

Au under the beam was observed but is not shown. At 150 °C dark spots are visible on some CdSe nanopyramids and a few larger particles with high contrast (top right part of Figure 3.14 b). The nanoparticles are less well ordered and shifted even more after heating to 200 °C. Instead of network formation by welding, large Au nanoparticles were formed.

This emphasizes further that the interaction with electrons in the TEM has a more complex influence on the migration of Au than only heat generation. Chen *et al.* drew a connection between the destruction of ligands by interactions of the sample with an electron beam and surface diffusion of Au [283]. They proposed that secondary electron generation in Au is responsible for the desorption of ligands rather than the generation of heat. Instead of ripening this process causes surface diffusion which leads to the formation of "necks" between nanoparticles. This explains the noticed faster diffusion of Au in shell samples re-dispersed with additional ligands after the deposition compared to non-treated samples. The outer layer of atoms becomes more mobile during destruction of ligands which entails further diffusion and induced crystallisation of more stable Au domains. An electron transfer that induces the desorption of ligands and thus favours atomic rearrangement of the meta-stable shell to dots also makes the earlier described effect of the reducing agent tetrabutylammonium borohydride comprehensible.

The deposition of Au onto CdSe nanopyramids has shown that the formation of an Au(0) based shell is influenced by the oxidation state of the metal precursor and the peculiar surface faceting of the seeds. An Au precursor with a lower oxidation state favours the formation of defined oligomeric hybrid nanoparticles in a reduction reaction controlled deposition on the surface of the seed.

Instabilities against the electron beam can be understood as arising from a reduction of the small content of Au(I) components on the one hand and from the destruction of ligands bound to the surface on the other. Both effects can increase the mobility of surface atoms and lead to rearrangement coupled with induced crystallisation of Au. As ligands are necessary to stabilise the nanoparticles and Au is more susceptible to diffusion than silver [284], for instance, Au-CdSe nanostructures seem to be prone to changes in their configuration especially in applications involving electron flux. Furthermore, their annealing is related to ripening processes along the seed crystal [169], which could leave metal atoms inside the semiconductor and thus result in less defined structures. For these reasons it is desirable to find a more stable alternative for elemental studies of hybrid nanoparticle properties.

3.2.2 Reactions of CdSe nanopyramids with Ag, Pd and Pt

The reaction conditions were modified for the synthesis of hybrid nanoparticles with silver, palladium and platinum, owing to the solubility of the chloride precursors or a lack of deposition in the case of platinum. A single phase reaction in organic solution was chosen, since this should hinder ion exchange processes [248]. Following the idea to obtain a universal method for different metals, a straightforward recipe for the preparation of metal nanoparticles in toluene with oleylamine, acting both as ligand and reducing agent, was adapted [285]. Metal acetates or acetylacetonates were dissolved in toluene and oleylamine before CdSe nanopyramids were injected and left to react under toluene reflux for varied periods of time. In particular, 31 µmol metal salt and 1 mL (2.4 mmol) oleylamine were mixed with varied concentrations of CdSe nanopyramids in 12.5 or 25 mL of toluene. The metal to CdSe nanopyramid ratio was determined as described with Au in the section above.

3.2.2.1 Reaction with Ag(I)

In reactions with silver acetate quick ion exchange processes occurred, similar to the results with CdSe nanorods in literature [245]. After 10 minutes in a reaction with a Ag to CdSe ratio of 24, for example, no deposit but welding of the nanoparticles is visible in the micrograph in Figure 3.15.

CHAPTER 3. METAL-SEMICONDUCTOR HYBRID NANOPARTICLES

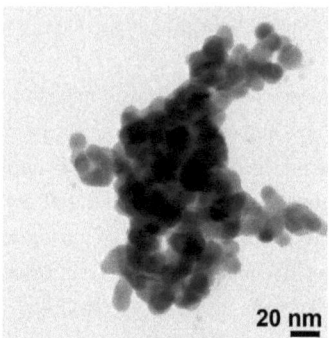

Figure 3.15: TEM micrograph of nanopyramids after ion exchange with Ag.

All Cd was exchanged according to EDX measurements and a stoichiometry of Ag to Se of 3:1 was recorded (the spectrum is located in Appendix A). Such a ratio corresponds to a mixed phase of α or β-Ag_2Se and Ag [286]. The diffraction pattern shown in Figure 3.16, anyhow, is not a simple combination of the orthorhombic α- or the cubic β-phase and Ag[15]. Neither is it similar to the tetragonal structure reported in [245]. A combination of mixed phases in the sample seems reasonable, since the transition temperature of 120 °C between them is close to the reaction temperature [286]. It is possible that nanoparticles in the same sample adopt different crystal structures. Elemental silver may form a covering or welding layer which is not as easily detected as in the above referred case of Au due to a lower contrast. The general result of the developed reaction with silver acetate is that an ion exchange could not be prevented under the given conditions, even with the reducing agent oleylamine. A way to inhibit the process in rod-shaped nanoparticles was demonstrated by Bala *et al.*, who applied a biphasic protocol with Ag ions dissolved in the aqueous layer and obtained metallic tips [162]. Such an approach was out of question for the current work with respect to possible oxidation reactions of the nanoparticles upon intensive contact with water. This would for instance deteriorate electrical properties of the nanoparticles.

---

[15] The assignment of α to the low temperature orthorhombic phase stems from reference [286], in some sources α and β are applied *vice versa*.

3.2 Metal-CdSe nanopyramid hybrid structures - deposition and ion exchange

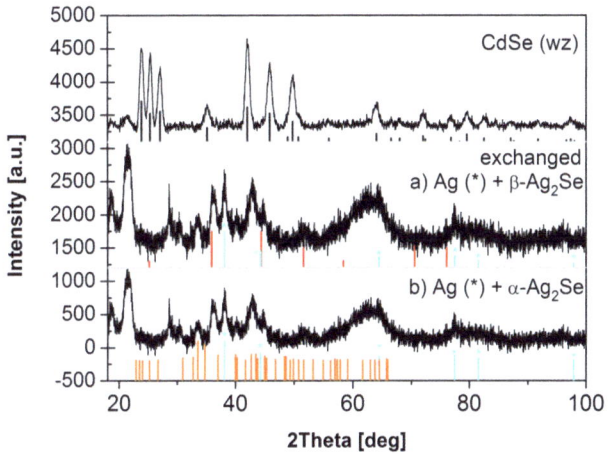

Figure 3.16: Powder X-ray patterns of CdSe nanopyramids and the product of ion exchange with Ag at a ratio of Ag/CdSe = 24. For a composition of Ag to Se of 3:1 mixed mixtures of Ag and two Ag$_2$Se phases are possible. a) Is a combination with the cubic β-Ag$_2$Se, which is thermodynamically more stable, and (b) is a combination with α-Ag$_2$Se, which is stable at room temperature. Reference data CdSe: JCPDS # 00-008-0459; Ag: JCPDS # 00-004-0783 [234]; α-Ag$_2$Se: JCPDS # 00-024-1041 [256]; (β-Ag$_2$Se): JCPDS # 00-027-0619 [287].

3.2.2.2 Reactions with Pd(II)

The activation barrier for an exchange of Cd(II) for Pd(II) in CdSe is higher than for Ag(I), since the contraction of the unit cell $\Delta V/V$ is substantially larger for evolving selenide phases with Pd (-0.29 to -0.30) than with Ag (0.06) [248]. Nevertheless, ion exchange processes occurred in the experiments, albeit significantly slower than in the case of Ag. The progression of a reaction of nanopyramids, synthesised with DCE, with palladium acetate at a Pd/CdSe ratio of 13 is shown in Figure 3.17. Increasing areas with dark contrast correspond to a proceeding exchange between the cations. The process shows similarities to the mechanism found with CdS nanorods and Ag where AgS$_2$ nucleated non-selectively on the surface of CdS [288]. There, with increasing size of these domains the growing lattice strain finally lead to their ripening inwards into the seed structure and the formation of segments along the rod as interfaces were reduced.

CHAPTER 3. METAL-SEMICONDUCTOR HYBRID NANOPARTICLES

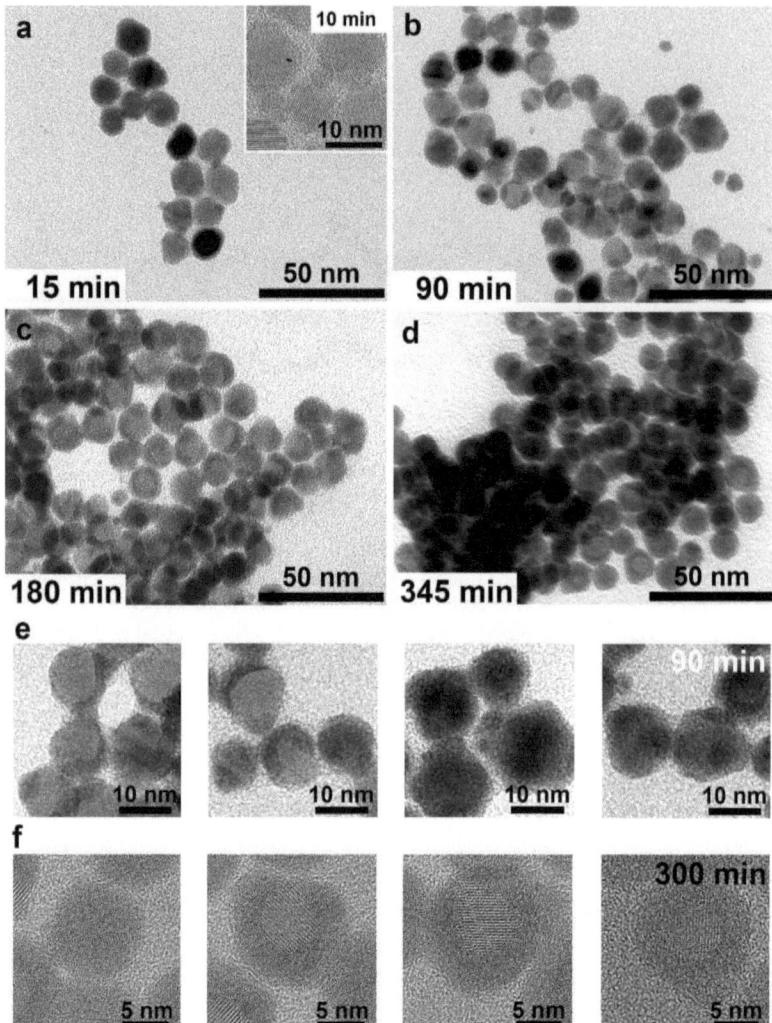

Figure 3.17: (a - d) TEM micrographs of CdSe nanopyramids at different stages of ion exchange between Cd(II) and Pd(II). (e, f) Higher magnification of micrographs at different spots after 90 and 300 minutes. Smaller nanoparticles showed complete exchange after 300 minutes, while the shell structure in some of the larger ones was slightly open to one side.

3.2 Metal-CdSe nanopyramid hybrid structures - deposition and ion exchange

In the current samples exchanged regions grew along the surface until a core-shell structure emerged, in which no core was visible in smaller nanoparticles as in Figure 3.17 f. Together with the exchange a small reduction of the diameter from 13.1 ×12.2 nm to 13.1×11.4 nm was noted, indicating a contraction of the crystal lattice. Elemental analysis of the original CdSe nanopyramids and samples reacted with Pd(II) for 10 and 300 minutes by EDX revealed a successive loss of Cd with a concomitant gain in Pd. After 300 minutes Pd was in excess to both other elements, indicating that the exchange for Cd happened at a ratio different from 1:1.

Table 3.3: Elemental composition of CdSe nanopyramids (DCE) reacted with Pd precursor.

Sample	Composition atom%				
	Cd	Se	Pd	ratio Cd/Se	ratio Pd/Se
CdSe	42.4	57.6	-	0.714	-
Pd-CdSe 10 min	39.3	57.3	3.3	0.667	0.06
Pd-CdSe 300 min	7.7	39.4	53.0	0.17	1.4

A complete exchange in all nanoparticles was tried to be achieved by longer reaction times, higher concentrations of Pd precursor in the beginning or injected later on. Despite various attempts, this was prevented by unresolved difficulties with irreversible precipitation of the nanoparticles during the reaction. The precipitation occurred mostly at a stage where Pd was visible around the surface in TEM images but in some cases even within the first hour. A reason for this might be a changing quality of oleylamine batches resulting in reduced stabilisation. Anyhow, the use of other amines (dodecylamine, hexadecylamine), palladium acetylacetonate instead of the acetate and additional injections of ligands at the reaction temperature had no significant effect on the sample stability or only lead to a temporary stabilisation and retardation of the shell formation (more oleylamine, hexadecylamine). The post-synthetic addition of oleylamine or dodecanthiol before purification provided no improvement.

A similar problem occurred with nanopyramids prepared with 1-chlorooctadecane. In this case network-like structures connecting several nanoparticles formed during 5 hours of reaction at a Pd/CdSe ratio of 16 (Figure 3.18 a). A possible but speculative explanation for this unexpected behaviour may be a decrease in the polarity of the reaction solution caused by long-chain alkyl residues attached to the nanoparticles from the initial synthesis. A more hydrophobic solution might then force the nanoparticles and precursor closer together. In the high resolution image in Figure 3.18 b the crystal lattice of the remaining CdSe core is well recognisable. The surrounding shell showed an amorphous structure.

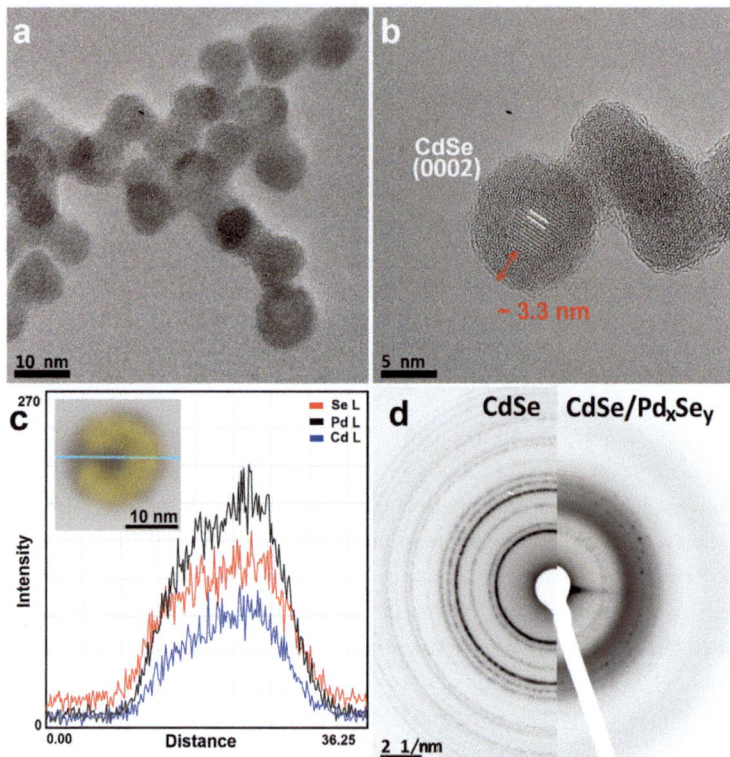

Figure 3.18: (a, b) TEM micrographs, (c) EDX line scan with inset containing the STEM micrograph in false colours and (d) electron diffraction with inverted colours of partially exchanged CdSe/Pd$_x$Se$_y$ nanoparticles obtained from nanopyramids prepared with 1-chlorooctadecane. The atomic composition of the sample in % was Se: 45.6, Pd: 49.5 and Cd: 4.9. The false yellow colour in the inset with inverted colours in (c) indicates regions of higher contrast and thus an incomplete shell.

3.2 Metal-CdSe nanopyramid hybrid structures - deposition and ion exchange

The signals of Cd and Pd in EDX are so close to each other that a locally resolved analysis of both elements occurring in one sample is difficult. This counts for attempts to identify whether Pd is mixed with Se in domains in early reaction stages and for an analysis of a core-shell structure. In the line scan in Figure 3.18, Cd is less present towards the surface of the nanoparticle, especially towards the right hand side, which is darker in contrast in the STEM image with inverted colours (marked yellow for better visibility). Apart from the rings belonging to the CdSe core, the electron diffraction pattern only exhibits broad features. Together with the amorphous shell visible in TEM micrographs this may be explained by a metastable state originating from the incomplete cation exchange. If nanoparticles are comparably large, the reaction zone, where original and new cations are present simultaneously, is smaller than the particle diameter [245]. Since this zone contains a certain degree of disorder and the exchange reaction happens slowly, it is possibly the reason for the undefined structure of the shell. With a composition of Se/Pd/Cd = 45.6:49.5:4.9 determined by EDX, the Pd/Se ratio is 1.08. This value is close to the 1.1 ratio that would be found with the cubic $Pd_{17}Se_{15}$ structure and indicates that the exchange may proceed towards this stoichiometry. The same phase has earlier been formed through the reaction of spherical CdSe nanoparticles with Pd precursor in a biphasic reaction solution [248].

The optical properties of the samples reflect the increasingly undefined structure of the nanoparticles (Figure 3.19). With proceeding reaction time the absorption bands of CdSe shift to shorter wavelengths and become flatter or even featureless towards the end of the reaction. Also with an extension of the measuring regime to longer wavelengths (1300 nm) no newly appearing bands were recorded. Emission spectra measured from 550 to 1700 nm were also without reliable features in all of the samples with Pd.

With the occurring ion exchange processes the system presents an opportunity for the preparation of semiconductor core-shell structures but is not suitable to further examine electrical properties within the aims of this work, for instance. In this study defined metal-semiconductor hybrid nanoparticles with separate phases are more advantageous.

CHAPTER 3. METAL-SEMICONDUCTOR HYBRID NANOPARTICLES

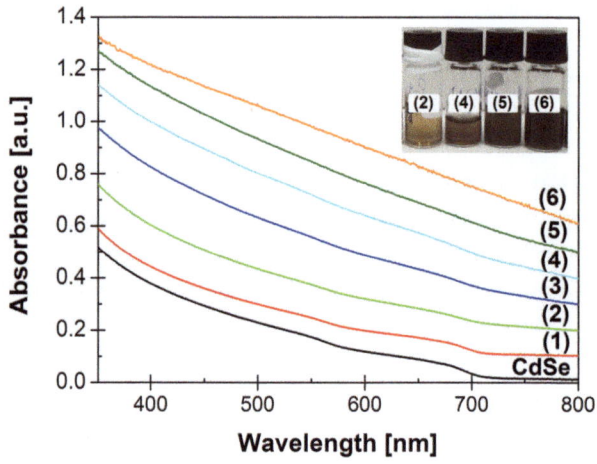

Figure 3.19: Absorbance spectra of samples from a reaction of CdSe nanopyramids with Pd precursor (Pd/CdSe: 26). The numbers stand for reaction times: (1) 10 min, (2) 60 min, (3) 90 min, (4) 120 min, (5) 180 min, (6) 300 min. The photograph in the inset illustrates the darkening of the dispersion with time.

3.2.2.3 Reactions with Pt(II)

Such stable oligomeric hybrid nanoparticles could be obtained by depositing Pt onto CdSe nanopyramids. With platinum acetylacetonate $(Pt(acac)_2)$[16] Pt deposits could be tuned from sub-nanometre clusters to above 3 nm domains simply by adapting the reaction time at a Pt/CdSe ratio of 24 (ratio mol Pt:mol CdSe nanoparticles = 18500:1). There was no indication of ion exchange processes. A reason for this could be the necessary contraction of the unit cell of $\Delta V/V$ = -0.27 for Pt_4Se_5 [248] in combination with the high reduction potential of 1.18 V.

Micrographs of samples at different reaction times are shown in Figure 3.20. Size tunability is an important synthetic feature in regard to applications, since size dependent properties, for instance of photocatalytic [213] or (opto)electronic nature, may be investigated. The slow growth provides conditions for a high crystallinity of the metal domains. During the growth CdSe nanoparticles of the sample shown shrank slightly after 24 hours, from $12.4 \pm 2.2 \times 13.1 \pm 2.4$ nm to a diameter of 10.8 ± 1.9 nm (the c-axis was not well distinguishable after 24 hours). A decrease of the rod length was also observed in the deposition

---
[16] The acetylacetonate is more accessible than the acetate and thus commonly employed as Pt source.

3.2 Metal-CdSe nanopyramid hybrid structures - deposition and ion exchange

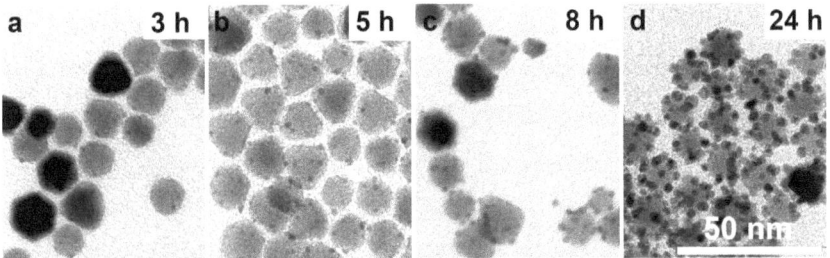

Figure 3.20: Pt-CdSe hybrid nanoparticle samples taken from a synthesis with Pt/CdSe = 23 after different reaction times. The diameter of Pt domains increases with time: (a) 1.0 ± 0.3 nm, (b) 1.5 ± 0.4 nm, (c) 1.8 ± 0.3 nm and (d) 3.3 ± 0.4 nm.

of Pt onto CdS in water [227], indicating a reduction mechanism under participation of CdSe surface atoms. Compared to the deposition of Au and Pd only minor changes in the atomic Cd/Se ratios are observed. Table 3.4 lists elemental compositions of the intial CdSe nanopyramids and samples after 1 and 24 hours at 110 °C as well as after 24 hours at room temperature.

Table 3.4: Atomic composition of CdSe and CdSe-Pt samples determined by EDX.

Sample	Composition [atom%]			ratio Cd/Se
	Cd	Se	Pt	
CdSe	43.8	56.2	-	0.769
CdSe-Pt 1 h 110 °C	41.5	58.0	0.5	0.714
CdSe-Pt 24 h 110 °C	32.1	46.5	21.4	0.714
CdSe-Pt 24 h 22 °C	45.3	54.7	-	0.833

At room temperature no platinum was detected. The lack of deposition is also evident from the different colours of the samples after 24 hours shown in a digital photograph in Figure 3.21. Absorption and emission spectra confirm these findings. In accordance with other Pt-CdSe hybrid nanoparticles described earlier, the emission is efficiently quenched in the high temperature reaction with deposited Pt domains, while it is maintained at room temperature. In the absorption spectra tailing at higher wavelengths and an increase of absorbance towards lower wavelengths evolves with time. These effects are attributed to cross-interfacial electronic interactions and absorption of the platinum domains [194].

CHAPTER 3. METAL-SEMICONDUCTOR HYBRID NANOPARTICLES

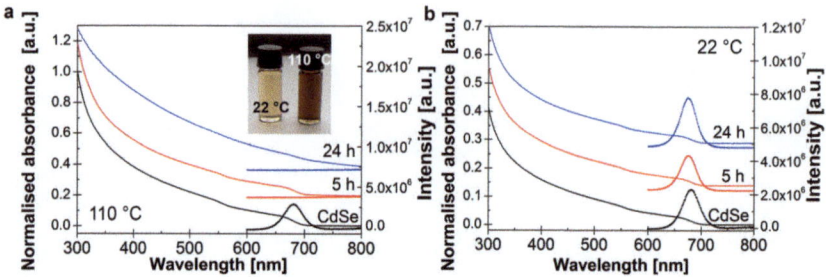

Figure 3.21: Absorbance and emission spectra of Pt-CdSe hybrid reactions conducted at (a) toluene reflux (110 °C) and (b) room temperature (22 °C). The inset in (a) shows a photographical image of the two reaction mixtures after 24 h. The absorbance spectra were normalised with the value at 300 nm. The corresponding emission in (b) was multiplied with the same factor.

From the results it is obvious that the reduction and deposition of Pt require thermal activation, which implies that, similar to the original metal nanoparticle synthesis, oleylamine plays a major role in the reduction of the precursor.

There is no epitaxy between the CdSe and Pt lattices in the strict sense of that the deposited material continues in the orientation of the seed material. In many cases there is an angle between the lattice fringes as in the high resolution micrograph and insets in Figure 3.22. Similar as with dot-shaped Au-deposits interfacial contact with step dislocations is observed (compare Figure 3.5). The two-dimensionality of the projection in the micrographs hinders an analysis of the lattice orientations between CdSe and Pt free of doubt. Nevertheless, Pt nucleated on a variety of facets on the surface of CdSe. This comparatively low site selectivity may be caused by a slow reduction process and the increased temperature necessary for Pt deposition. In heated samples ligand dynamics of adsorption are increased compared to room temperature conditions, so that surface sites become temporarily uncapped and thus better accessible for metal ions. During TEM inspection no ripening of Pt domains was observed. The identity of the Pt domains is additionally verified by EDX mapping in the scanning TEM mode (Figure 3.23). Their crystallinity and cubic structure are well documented by selected area electron diffraction shown in Figure 3.22 and X-ray diffraction (the X-ray diffraction pattern is shown in Appendix A).

3.2 Metal-CdSe nanopyramid hybrid structures - deposition and ion exchange

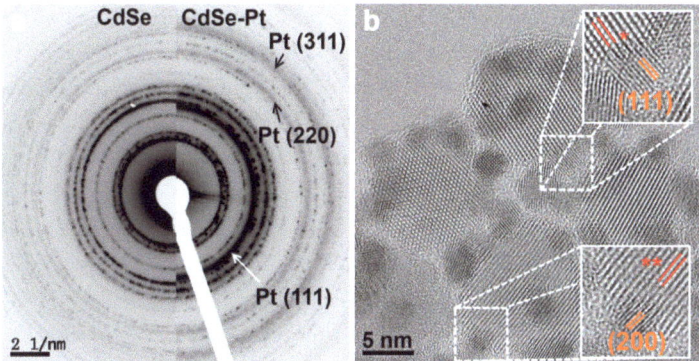

Figure 3.22: (a) Electron diffraction of CdSe before and after deposition of 3.3 nm Pt domains. The rings belonging to Pt are broadened due to various orientations towards the electron beam. (b) High resolution micrograph of the sample with two examples of interfaces in the insets. The marked lattice fringes are in the upper inset *CdSe (1010) and Pt (111) and in the lower inset **CdSe (0002) and Pt (200) planes.

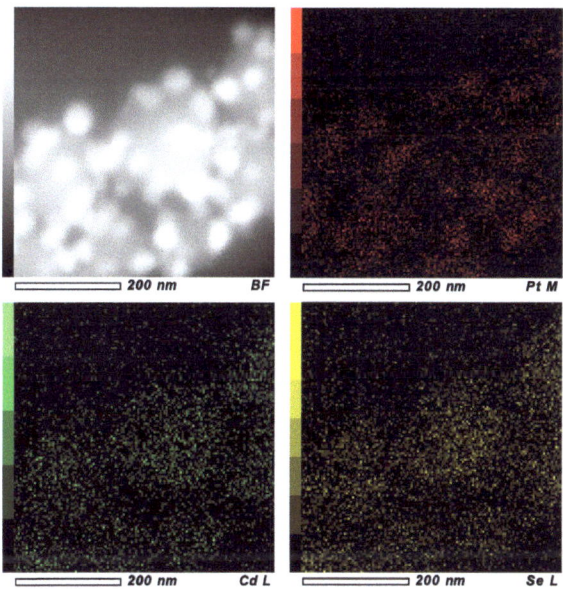

Figure 3.23: STEM micrograph and EDX map of Pt-CdSe hybrid nanoparticles after 24 h with Pt (upper right), Cd (lower left) and Se (lower right).

To further test the stability of the hybrid nanoparticles, *in situ* annealing inside the TEM was carried out. Micrographs recorded between 25 °C and 440 °C are compiled in Figure 3.24 and at higher resolution[17] in Figure 3.25. From 300 °C on, some of the nanoparticles began to form connections via Pt domains at 400 °C shuffling and a tendency towards the formation of chain-like structures occurred. A deformation of some Pt domains to cubic structures is visible with high resolution at 400 °C. Between 400 °C and 440 °C sublimation of the CdSe component set in. The weakly visible, frizzy pattern around and between the nanoparticles from 300 °C to 440 °C can be interpreted as a sign for carbonisation processes known to take place with platinum rich nanoparticles [289].

The above results show that the synthesised hybrid nanoparticles with Pt are stable and do not show temperature dependent intraparticle ripening processes towards less and bigger domains as reported for Au-CdSe nanorods [169]. In comparison with the standard synthesis of Pt-tipped CdS and CdSe nanorods less reagents and lower temperatures are necessary [163, 199] with the developed method. The size of the Pt domains can be easily tuned by varying the reaction time. These characteristics qualify the system for further application such as studies on electrical transport and potential use in (opto)electronic devices.

---

[17] New spots were chosen for each micrograph to reduce misinterpretations caused by interactions with the electron beam.

3.2 Metal-CdSe nanopyramid hybrid structures - deposition and ion exchange

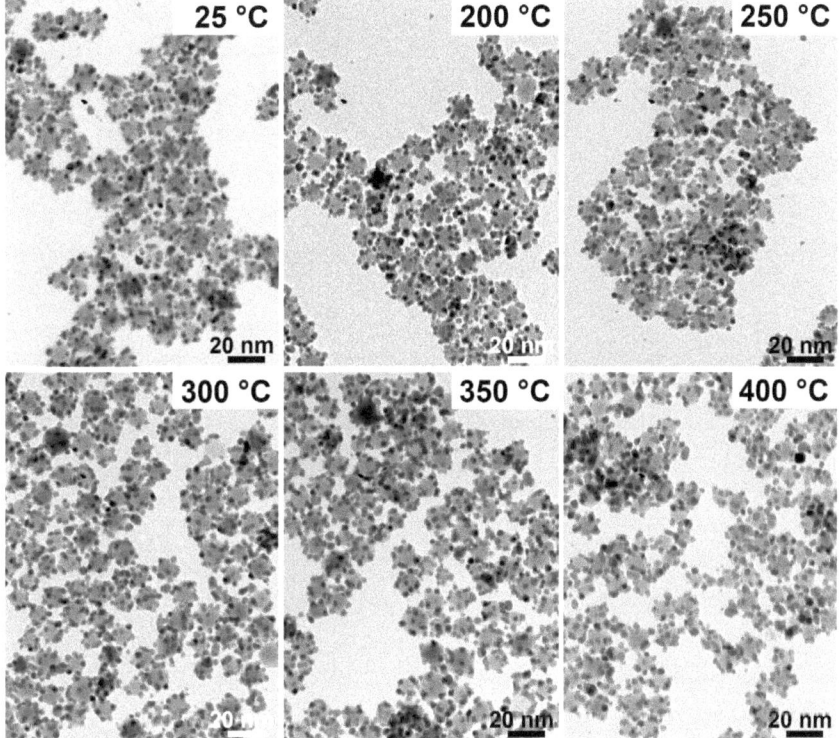

Figure 3.24: TEM micrographs of different spots on a Pt-CdSe nanoparticle sample after 24 hours *in situ* annealed in the TEM at different temperatures.

CHAPTER 3. METAL-SEMICONDUCTOR HYBRID NANOPARTICLES

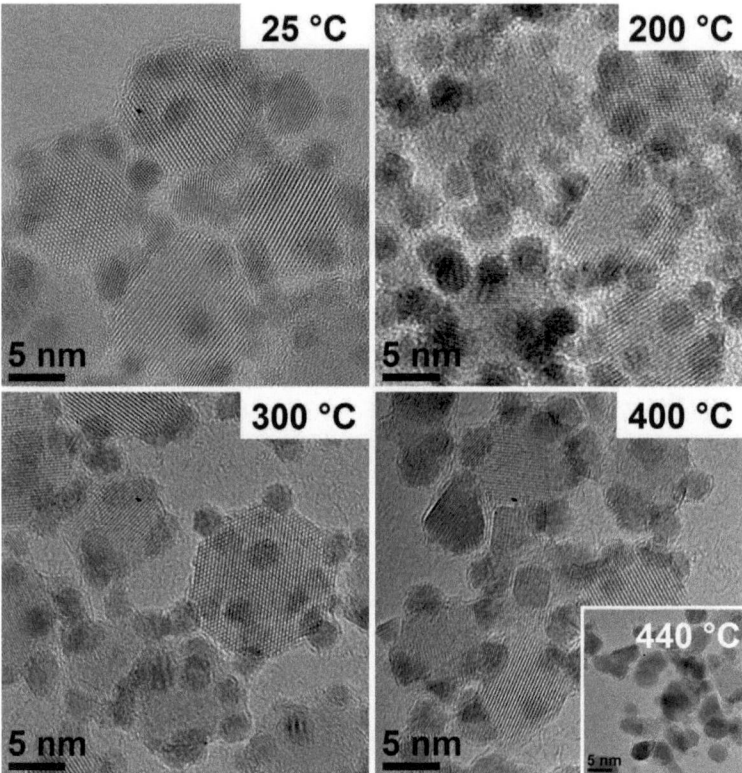

Figure 3.25: High resolution micrographs of Pt-CdSe nanoparticles *in situ* annealed in the TEM at different temperatures.

3.3 Conclusions

With the aim of obtaining oligomeric metal-CdSe nanopyramid structures, the deposition of four different metals onto CdSe nanopyramids was examined: Au, Ag, Pd and Pt. Seeded growth methods in organic solution were applied. With Au, the deliberate formation of an unstable shell or crystalline spherical domains was examined. Ion exchange processes characterised the developed reaction of the nanopyramids with Ag and Pd precursors, while stable and size tunable spherical domains were obtained with Pt. Even though the precursors were not completely comparable due to different oxidation states in addition to varied types and numbers of counterions, the observed reactions between different metal precursors and CdSe seem to be strongly influenced by the corresponding reduction potentials. These can be altered slightly by ligands which usually reduce the potentials and make a deposition less feasible [182]. In the line of the employed metals, Au with the highest reduction potential was most easily reduced, especially when added in oxidation state (I), followed by Pt(II) at elevated temperatures. Even though the mild reducing agent oleylamine was present and thermal energy was applied, the metal ions Pd(II) and Ag(I) with reduction potentials close to or below that of Se underwent ion exchange processes with CdSe seeds. The existence of the high number of crystallographically exposed reactive sites in CdSe nanopyramids did not affect this behaviour.

More detailed conclusions concerning Au are that the choice of the precursor, containing Au(I) or Au(III), directs the deposition of the metal onto either exposed crystal sites or the whole surface of the nanoparticles. The preferential coordination and reduction of Au by Se lead to a loss of Cd from the nanoparticles during deposition. The Au-Se interaction is the reason for the formation of a mostly elemental but unstable Au shell with pyramids that is not observed with rod-shaped nanoparticles, since the sloped facets can become Se rich by removal of Cd. An attempt to exploit this instability and form a network of Au-joined CdSe nanoparticles through annealing of a monolayer only resulted in the formation of large Au spheres separate from CdSe. The occurring instabilities of Au-CdSe hybrid nanoparticles against electron irradiation and the thermally induced ripening reported elsewhere lead to the conclusion that they lack the necessary stability in configuration for further studies such as electrical characterisation.

Reactions with silver lead to ion exchange processes and undefined structures, which were considered undesired for further investigations. A biphasic deposition suggested in reference [162] might be a solution to obtain spherical domains but bears the danger of oxidation of the nanoparticles due to intimate contact with water.

CHAPTER 3. METAL-SEMICONDUCTOR HYBRID NANOPARTICLES

CdSe nanoparticles subjected to Pd precursor underwent an ion exchange process to partially substituted core-shell structures, which precipitated irreversibly before the exchange was completed. The colloidal instability of the partially exchanged nanoparticles and difficulties with the removal of residues from the synthesis, gluing the nanoparticles together, disturbed further processing and a more detailed characterisation. The shell was of no defined crystallographic phase but if the so far unresolved difficulties with precipitation of the nanoparticles will be solved, these structures might be annealed to develop a distinct Pd_xSe_y phase. This might lead to interesting physical properties since some Pd-Se phases are superconducting [290] at low temperatures. With CdSe nanopyramids synthesised with 1-chlorooctadecane a tendency towards network formation or welding of the core-shell structures occurred. While these networks are not well suitable for electrical studies, they might be interesting for thermoelectric or catalytic applications.

Stable oligomeric hybrid nanoparticles were finally obtained with Pt. The domains could be varied from sub nanometre size to diameters of around 3.3 nm. The nanoparticles showed emission quenching which is a sign for efficient charge transfer from the semiconductor to the metal. No instabilities under influence of the TEM beam or ripening through thermal heating were observed. These characteristics make the nanoparticles suitable candidates for photocatalytic or electrical characterisation. Owing to the higher novelty of the research, studies on the latter provide broader perspectives.

4 Electrical transport in hybrid nanoparticle films

Nanoparticle solids and thin films own a large potential for electronic and optoelectronic devices as concisely reviewed by Talapin *et al.* [3]. Low material use and quantum size effects offer the possibility of creating efficient and compact structures, for instance for light generation (LEDs) [20, 291, 292], photovoltaics [18, 293] and sensor applications [294, 295]. A major advantage of nanoparticles in such applications is their solution processability, which allows for a variable and cost effective production of arrays via self assembly [25, 296], spin- and dip-coating as well as Langmuir-Blodgett assisted monolayer formation [3, 297]. Hybrid nanostructures consisting of metal and semiconductor components are often referred to as promising materials for optoelectronic devices. Regardless of this, the majority of studies on applications of such systems is related to charge transfer processes in solution and in particular to photocatalysis. For this reason, it is interesting to learn more about electrical transport in hybrid nanoparticle arrays. Especially since direct metallic contacts can significantly improve conductivity without the need for post-deposition treatments of semiconductors [13]. Metallic domains might even introduce effects such as Coulomb blockades at low temperatures [158]. In the present work, the electrical transport properties of Langmuir-Blodgett deposited CdSe and CdSe-Pt nanoparticles were examined to elucidate the effects of heterodeposited metal domains on conductivtiy and photoconductivity of semiconducting nanoparticle arrays. Pt-CdSe hybrids were chosen due to the tunability of the Pt domain size and their stability (see Chapter 3, Section 3.1.1). Furthermore, at 300 K the Schottky barrier between CdSe and Pt bulk material is nearly half of the barrier between CdSe and Au (0.37 and 0.7 V [298]), which is advantageous for the charge transfer between the metal and the semiconductor. There is a lack of models for transport in two-dimensional arrays of oligomeric hybrid nanoparticles of this kind. Hence, explanations for the obtained results will be sought in transport theories based on semiconductor and metal components combined with properties of metal-semiconductor contacts and experimental studies on single/network hybrid particles, as argued in the following.

CHAPTER 4. ELECTRICAL TRANSPORT IN HYBRID NANOPARTICLE FILMS

4.1 Electrical transport in nanoparticle arrays

Compared with macroscopic devices, nanoparticle arrays exhibit special characteristics due to size effects and related quantum physical phenomena. In the following, only the aspects relevant for the electrical transport in the presented systems will be explained, a broader and more detailed overview is provided by reviews dealing with nanoparticle applications [3, 299] and in books covering different aspects of nanoparticles [43, 44, 300]. Electrical transport in arrays of crystalline nanoparticles is often measured with two-probe electrode configurations (source and drain) such as the set-up displayed in Figure 4.1. Factors observed to contribute to the transport characteristics of such devices and the major aspects of their impact are:

- nanoparticle material (composition, size, doping) ⇒ electronic structure
- order, interparticle distance ⇒ barrier widths
- electrode - nanoparticle contact ⇒ charge injection, barrier height
- device geometry and film thickness ⇒ dimension of transport, number and length of possible paths

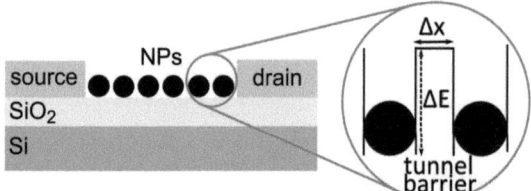

Figure 4.1: Schematic representation of a nanoparticle based source-drain device with enlarged depiction of a tunnel barrier at zero bias with depiction of the barrier height ΔE and width Δx (the prior is situated in energy space, the latter in position space).

The composition and the size of the nanoparticles determine the electronic structure and thus influence optical properties (see Chapter 2, Chapter 3) as well as the temperature dependence of the materials' conductivity. Nevertheless, grain boundaries and ligand filled gaps between nanoparticles in electronic devices have significant consequences for the mobility of charge carriers and the mechanisms of their transport. While the transport in bulk materials is dominated by conduction via electronic bands, three-dimensional coupling of

4.1 Electrical transport in nanoparticle arrays

wavefunctions to band-like states (minibands) only exists in solids consisting of quantum dot superlattices if the array fulfils strong and in some cases hard to achieve requirements. Among these are monodispersity of the nanoparticle shape and size (standard deviation $\sigma_{size} < 5\ \%$), formation of a long range order and small, material specific distances which enable coupling of wave functions between the nanoparticles [3, 70, 301, 302]. Most of the organic stabilizers employed in the synthesis of nanoparticles contain long hydrocarbon chains and reduce the electronic coupling by forming insulating barriers between deposited nanoparticles causing the latter to present localised states for the charge carriers. The width Δx and height ΔE of tunnel barriers between the nanoparticles as visualized in Figure 4.1 determine the degree of coupling between electron wave functions by contributing to the approximated tunnelling rate[18] Γ

$$\Gamma \approx exp\left[-2\left(\frac{2m^*\Delta E}{\hbar^2}\right)^{1/2}\Delta x\right] \quad (4.1)$$

with the effective mass of the electron m^* and the reduced Planck constant $\hbar$ [3]. The tunnelling rate scales exponentially with the interparticle distance, which is the reason why the packing of nanoparticles plays a decisive role for the performance of the devices. Methods employed to reduce interparticle distances include annealing and exchange of the stabilisers [3, 303, 304]. Desirable stabilisers may be amphiphils with shorter hydrocarbon chains than those employed during synthesis, sometimes also with cross-linking properties [296, 305, 306], aromatic compounds [296, 304, 307], molecular chalcogenide ligands [308] or atomic ligands (for example halides) [27, 28, 29, 32]. A new strategy is the exchange for ligands that decompose under heating and leave the surface uncapped [309, 310, 311].

The contact between electrode and nanoparticle array is thought to be another bottleneck for transport measurements. Variations of electrode materials indicated that the band alignment of the two materials determines the feasibility of charge carrier flow in and out of the sample [312]. Further details will be explained in Section 4.1.2.

In the following, different transport mechanisms relevant for processes in structures containing semiconductor and metal nanoparticles as well as hybrid nanoparticles will be summarised and applied to analyse the results.

---

[18] also named *transmission coefficient* [70] or *tunnelling transmission probability* [299]

CHAPTER 4. ELECTRICAL TRANSPORT IN HYBRID NANOPARTICLE FILMS

4.1.1 Transport mechanisms and photoconductivity

Due to low coupling across the interparticle barriers, most semiconductor and metal nanoparticle arrays electronically behave like non-crystalline or disorderd solids with localised states (Anderson-Mott insulators) [297, 299, 303, 313, 314, 315]. Thus, electronic transport is strongly limited by size dependent quantum mechanical properties and mainly based on thermally activated hopping processes between nanoparticles.

4.1.1.1 Coulomb blockade and single-electron transistor

For semiconductor nanoparticles in the confinement regime and small metal nanoparticles the capacitance C becomes a limiting factor for charge transport

$$E_{Ch} = \frac{q^2}{2C} = \frac{q^2}{4\pi\varepsilon\epsilon_0 d} \tag{4.2}$$

and charge quantisation may become relevant [316]. In Equation (4.2) q is the elementary charge, ε is the permittivity of the dielectric surrounding the particle, ϵ_0 is the vacuum permittivity and d is the diameter of the nanoparticle. At low temperatures and small applied voltages, the charging energy E_{Ch} may outnumber the thermal energy $k_B T$ resulting in blocking of charge transport from electrode to electrode through a nanoparticle mantled by tunnel barriers, as depicted schematically in Figure 4.2. Such a Coulomb blockade can be overcome by a higher applied voltage, so that it is mainly observed for small particles at low temperatures and at close to zero applied voltage. In a single-electron transistor, a nanoparticle is located between a source, a drain and a gate electrode. Its limited, gate-dependent capacitance can be employed to accurately tune the number of

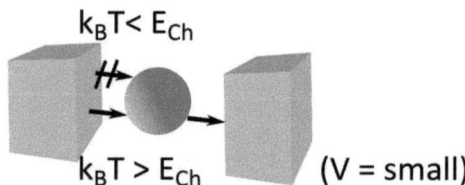

Figure 4.2: A Coulomb blockade between electrodes and a nanoparticle island occurs at low voltages V when the energy for the addition of a further charge E_{Ch} is higher than the thermal energy $k_B T$.

electrons on the particle and create a finite conductance. For further reading on Coulomb blockades and single-electron tunnelling processes references [190, 316, 317, 318, 319, 320] are recommended.

4.1.1.2 Transport mechanisms in non-crystalline materials

There are three basic mechanisms for electron transport in non-crystalline semiconductors: (i) transport at the lower edge of the conduction band with temporary stays in trap states (CB), (ii) nearest neighbour hopping between localized trap states (NNH) and (iii) variable range hopping (VRH), as illustrated in Figure 4.3 [321]. Their occurrence is related to the temperature T, in which (i) requires the highest thermal energy and (iii) the lowest.

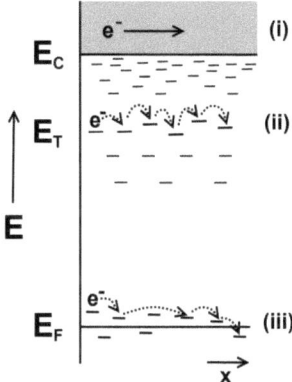

Figure 4.3: Electron transport mechanisms in disordered semiconductors with E_c lower band edge conduction band, E_T tail states and E_F Fermi energy [321].

Trap state influenced conduction band transport (i)

A general temperature dependency of the conductivity σ of n-type semiconductors above room temperature dominated by mechanism (i) can be formulated as

$$\sigma = \sigma_0 exp\left[\frac{-\Delta E}{k_B T}\right]. \tag{4.3}$$

Here, σ_0 is a pre-exponential factor, ΔE is the energetic difference between the lowest level of the conduction band E_C and the Fermi level E_F (for p-type semiconductors this would be the difference between the Fermi level and the highest level of the valence band), k_B is the Boltzmann constant.

Mechanisms (ii) and (iii) dominate at reduced temperatures and were reported to occur in semiconductor and metal nanoparticle arrays alike.

Nearest neighbour hopping (ii)
Below room temperature electrons located in trap states are often not injected into the conduction band but move by thermally activated transitions between localized states. This process, first reported by Miller and Abrahams [322], is referred to as *nearest neighbour hopping* and can be formulated as a simple Arrhenius-like temperature dependence of the conductivity [300]

$$\sigma(T) = \sigma_0 exp\left[\frac{-E_A}{k_B T}\right]. \tag{4.4}$$

The activation energy E_A is proportional to the distance between nearest neighbours and the Coulomb charging energy of a single particle, the conductivity increases with the density of hopping centres [321, 323, 324]. For three-dimensional CdSe nanoparticle arrays simple activated hopping was observed in the temperature range between 180 and 300 K with E_A being nanoparticle size independent [312]. In three-dimensional PbSe nanoparticle samples on the other hand, a dependency of the hopping transport on the particle size was registered [305]. Effects of size distribution were negligible as percolation paths formed between larger nanoparticles. Hopping transport in two-dimensional arrays of metallic cobalt-platinum nanoparticles was shown to follow a Neugebauer-Webb relation similar to Equation (4.4); the activation energy there is dominated by the charging energy of the single nanoparticles and is thus size-dependent [297, 325, 326, 327].

Variable range hopping (iii)
VRH will occur when hopping to the nearest neighbour is energetically less favourable than hopping to a state closer in energy but further in distance. This means that the hopping probability in VRH depends both on the spatial and the energetic separation of two states at a constant temperature. The temperature of transition from nearest neighbour to VRH is array specific. The mechanism was described for granular metal films by Abeles [328]. Mott and Efros/Shlovskii (ES-VRH) related conductivity and temperature with the dimension D of the transport in disordered solids with

$$\sigma \propto exp\left[-(T_0/T)^{(1/(D+1))}\right], \tag{4.5}$$

4.1 Electrical transport in nanoparticle arrays

in which ES-VRH takes Coulomb interactions between the electrons into account [329, 330]. While three-dimensional arrays tend to follow relations with $D = 1$ (Abeles, ES-VRH), two-dimensional systems comply with $D = 2$ (Mott) [295]. In three-dimensional CdSe nanoparticle arrays interconnected through 1,4-phenylenediamine, VRH with $D = 1$ was reported for the temperature region of 4 - 120 K and found to strongly depend on the electric field [331].

4.1.1.3 Photoconductivity

Excitons photogenerated in a semiconducting material can be separated by an applied field to obtain a photocurrent. In semiconductor nanoparticle/quantum dot arrays, photoconductivity was shown to compete with radiative and non-radiative recombination processes (photoluminescence, recombination in trap states within the band gap) and to be wavelength dependent, in compliance with the optical absorbance spectra of the nanoparticles [312, 332]. The generated charges tunnel between the nanoparticles, influenced by the charging energy and the barrier widths.

Typical effects observed in organic ligand capped photoconductive CdSe arrays (three- and two-dimensional) are a linear increase of photoconductivity with illumination intensity (dissociation of excitons is the rate limiting step), persistent photoconductivity after switching off the light source and current quenching due to surface trapping of charges [312, 332, 333]. The latter two are related to the accumulation of positive charges in the array, which at least partially depends on the distance between the electrodes or rather the applied field [132, 334]. It was argued that electrons generated in the film close to the drain electrode are collected, while the holes cannot be transferred to the source electrode due to a low electric field[19].

4.1.2 Electrical transport in metal-semiconductor hybrid nanoparticles

Studies on electrical transport in metal-semiconductor hybrid nanoparticles are still rare and the increased degree of complexity of the crystal composition and morphology present a challenge when it comes to analysing the observed characteristics. Similar to their optical properties the electrical response of semiconductor-metal hybrid structures is expected to be determined by electronic interactions across the interface and mechanistic considerations are often based on macroscopic models.

---

[19] In reference [132] the gate electrode beneath the array was set to zero bias.

4.1.2.1 Macroscopic metal-semiconductor contacts

When a macroscopic metal (Me) and a semiconductor (SC) are brought into direct contact, an electrostatic barrier and a space charge region form in the semiconductor. The electron flow depends on the work functions Φ of the materials. In a metal-n-type semiconductor combination electrons can flow towards the metallic compound ($\Phi_{Me} > \Phi_N$) resulting in a positively charged space-charge region (depletion layer, DL) and a Schottky barrier with barrier height Φ_{SB} (Figure 4.4 a and b). Alternatively, the charges accumulate in the space charge region ($\Phi_{Me} < \Phi_N$) thus turning it into an accumulation layer [179]. In the semiconductor region close to the interface, surface states may form inside the band gap. These so-called metal-induced gap states (MIGS) generate a local charge in the semiconductor which arouses an image charge in the metal. Due to the resulting electrostatic potential the semiconductor bands bend close to the interface (within 1 nm) and align the Fermi level to reach charge neutrality [335, 336]. Depending on the semiconductor this can result in Fermi level pinning and an indifference towards the work function of the metal.

In general, there are five contributions to charge transport at metal-semiconductor junctions [298]. Electrons can (i) surpass the barrier by thermal activation (thermionic emission relevant at around room temperature), (ii) tunnel through the barrier or (iii) recombine with holes in the space charge region. Additionally, electrons diffuse into the depletion layer (iv) and holes from the metal diffuse into the neutral region of the semiconductor where they recombine with electrons (v).

In devices with nanoparticles there are two types of metal-semiconductor contacts to consider. One is the contact between the sample and the metallic electrode, the other one is the intrinsic contact between the components of the hybrid nanoparticles. In an electrode set up with n-type semiconducting nanoparticles such as CdSe in between, electrons have to cross a Schottky barrier and possibly even tunnel through a layer of organic ligands to reach the nanoparticles. For this reason, the relative positions of energetic levels in both materials may have an influence on measured dark currents. Charge injection from Au electrodes into three dimensional CdSe nanoparticle arrays, for instance, are considered unfavourable and dark currents reported are in the regime of the electronic noise [303, 306, 312]. A positive side effect of this is that currents measured under illumination with Au electrodes only stem from photogenerated charge carriers.

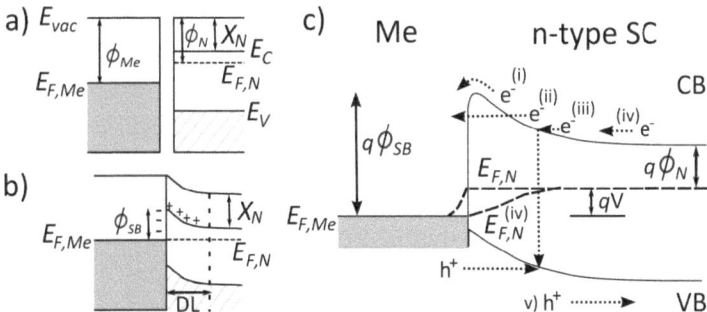

Figure 4.4: (a) Band structures of metal (Me) and n-type semiconductor (N) before and (b) after contact and equilibration. E_{vac}: Vacuum energy level; $E_{F,Me/N}$: Fermi levels; $\Phi_{Me/N}$: work functions; Φ_{SB}: Schottky barrier height; X_N: electron affinity of the semiconductor; DL: depletion layer; q: elementary charge; V: applied voltage. c) Transport mechanisms at metal-semiconductor junctions under forward bias. The occurring processes are (i) thermionic emission, (ii) tunnelling, (iii) recombination and diffusion ((iv),(v)). The barrier is lowered under an applied field. Under backward bias the processes are reversed [298].

4.1.2.2 Metal-semiconductor nanocontacts

Macroscopic space-charge regions extend over tens to hundreds of nanometres, thus often spanning more than the size of a nanoparticle [337]. Nevertheless, the concept of Schottky barriers is often applied to nanocontacts. Rare examples of long-distance band bending with depletion layers of bulk dimensions in one- or two-dimensional nanostructures are based on nanotubes or graphene, or else involve doping of the semiconductor [335]. Experimental and theoretical studies on single semiconductor rods and nanowires with deposited metal nanocontacts as well as thin film models of CdSe and Au layers revealed the existence of subgap states near the interface [282, 337, 338, 339, 340]. The occurrence of metal induced gap states in the centre of the semiconductor, which can lead to more metallic behaviour of the particle, seems to be strongly size and model dependent and remains a matter of discussion. In a scanning tunnelling microscopy examination of nanodumbbells with 40 nm CdSe rods the Au parts exhibited metallic conductance with Coulomb blockade effect, while an electronic gap structure with midgap states close to the interface and an unaffected band structure in the centre were found in the semiconducting part [337]. For such single nanodumbbell structures of Au tipped CdSe nanorods a five to six orders of magnitude increase in conductance was observed compared with bare CdSe nanorods [13]. A major impact on the increased conductivity was ascribed to the removal of the organic

CHAPTER 4. ELECTRICAL TRANSPORT IN HYBRID NANOPARTICLE FILMS

capping layer from CdSe by the Au deposition. Charge injection from the electrode was found to occur via tunnelling at low temperatures and thermionic emission above 250 K suggesting charge transport across a Schottky barrier. Krahne and co-workers studied the electrical properties of networks of CdSe nanorods decorated with several solution grown Au domains which were later welded to alternating metal-semiconductor structures by annealing [158, 341, 342]. The current-voltage characteristics of both single nanostructures and networks at around room temperature were described by thermionic emission processes. In the corresponding Schottky-Richardson relation for the current across a Schottky barrier under reverse bias conditions

$$I = AR^*T^2 exp\left[\frac{-q\Phi_{BE}}{k_BT}\right] \quad (4.6)$$

the charge transfer from metal to semiconductor is the limiting factor [13, 158, 298]. Contributions to the current are the electrically active contact area A, the effective Richardson constant R^*, elementary charge q, the effective barrier height Φ_{BE} and the thermal energy.

Schottky barrier controlled transport is not the only model that explains electrical properties of hybrid nanoparticle arrays. Talapin and co-workers examined transport phenomena in core-shell structures of Au and PbS as well as magnetic FePt cores with semiconducting CdSe, CdS, PbSe and PbS shells [165, 343, 344]. While they observed charge transfer from PbS to Au and thereby increased hole conductivity, core-shell structures with FePt showed no signs for charge trapping on the metal core and possible Fermi level pinning. On the contrary, the group proposed electron tunnelling processes between the metallic cores and suggested that the electronic properties of multicomponent nanostructures resemble those of binary superlattices.

Photoconductivity

While little is known about electronic transport in hybrid nanoparticles, even less information is available on photocurrents in such systems. To the best of the author's knowledge, there are only three studies including this aspect. Two of them are based on CdSe-Au systems. In networks of CdSe nanorods decorated with several Au domains similar to the above mentioned ones no photoresponse was observed [342]. Only after annealing where Au migrated from the side to tip positions on the nanorods a difference between dark and illuminated devices was detected. An enhancement of photocurrent was observed in CdSe nanowires when they were decorated with Au domains [345]. According to scanning photocurrent microscopy results a photocurrent was generated throughout the wire and the enhancement was attributed to Schottky fields that promote exciton dissociation and light scattering on Au. The third study showed that the presence of an Au core inside PbS nanoparticles permitted the generation of enhanced photocurrent instead of promoting non-radiative recombination [165].

4.2 Electrical transport in CdSe and Pt-CdSe nanoparticle devices

With respect to possible (opto)electronic applications of oligomeric metal-semiconductor hybrid nanoparticles, CdSe and Pt-CdSe nanoparticle samples were assembled into monolayers and their direct current characteristics were measured. This happened under vacuum in dark and illuminated conditions and at varied temperatures. The devices were prepared by assembling purified nanoparticles on a diethyleneglycol subphase by the Langmuir-Blodgett technique and depositing them onto Si wafers with a SiO_2 layer of 300 nm and interdigitated Au electrodes (distance between fingers: 0.41-0.45 µm), thereby following a procedure described in [132] with minor modifications (for details on the purification and Langmuir-Blodgett deposition see Chapter 5). The Langmuir-Blodgett technique offers the possibility to controllably prepare highly-ordered monolayers of nanoparticles [289, 297, 327, 346]. Interdigitated array electrodes as visible in Figure 4.5 allow for measurements over a comparatively large film area, average over a high number of paths, increase the currents and thus the signal to noise ratios measured for weakly conducting semiconductor samples. For this reason, such electrodes are ideal for pioneering qualitative measurements of Pt-CdSe hybrid nanoparticle samples.

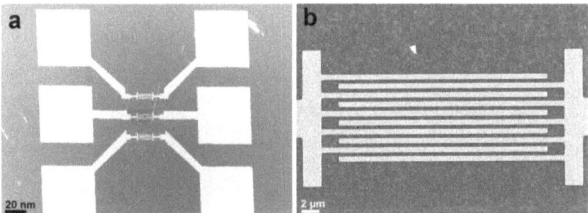

Figure 4.5: (a) Electrode structure with contact pads and (b) close up of its interdigitated area.

CHAPTER 4. ELECTRICAL TRANSPORT IN HYBRID NANOPARTICLE FILMS

Three types of nanoparticle arrays were examined: CdSe nanopyramids, CdSe nanopyramids with 3.2 nm sized Pt domains and CdSe nanopyramids with 1.7 nm sized Pt domains. CdSe nanopyramid monolayers have been examined by Cai on Si/SiO_2 devices with bar electrodes facing each other with the shorter axis and separated by distances of 0.6 to 2 µm [132]. The CdSe experiments presented here serve the purpose of verifying the observed characteristics with the employed elecrode geometry and to act as blank experiments for qualitative comparisons of room temperature measurements. The results of all samples will be discussed separately, since the density of nanoparticles in the arrays differs and the impression of direct comparability shall be omitted. Nevertheless, qualitative comparisons will be made and common conclusions will be drawn. The insights obtained are meant to serve as guidelines for future studies with the present and similar systems which will include the optimisation of sample preparation and detailed mechanistic examinations of transport processes.

4.2.1 Electrical transport through pyramidal CdSe nanoparticles

For the CdSe nanoparticle arrays, hexagonal dipyramidal CdSe nanoparticles with dimensions of 11.6 ± 1.4 nm × 10.6 ± 1.3 nm, prepared with 1,2-dichloroethan (see Chapters 2 and 3), were employed. The varying orientation of pyramidal particles in the film (some pyramids stand on their basal plane while others lie on the side) hinders the formation of highly-ordered layers and the determination of a uniform interparticle distance, for example by small angle X-ray scattering. The average center-to-center distance of the particles was 10.8 ± 1.3 nm with shortest distances between particles of around 1 nm caused by the alkyl-chain ligands [296]. Compression of the film lead to small areas with a second layer of particles. Due to the island-like distribution of these double layers and an otherwise dense monolayer underneath, the influence of the upper nanoparticles is assumed to be negligible. The data obtained by electrical characterisation of CdSe nanopyramid devices is summarised in Figure 4.7. Dark currents of CdSe devices measured in vacuum and at room temperature remained in the range of the noise level. This complies with the low dark currents reported for CdSe nanoparticle arrays contacted with Au electrodes [303, 306, 312].

Illumination of the device with the light of a CCD camera (150 W white light at 21 V) induced a photocurrent of roughly one order of magnitude higher than the dark current at 10 V lateral voltage. The photocurrent strongly depends on the applied voltage because the electric field ionizes the generated excitons and helps the charges to tunnel through the nanoparticles [332]. A hysteresis, which is characteristic for CdSe films, was observed

4.2 Electrical transport in CdSe and Pt-CdSe nanoparticle devices

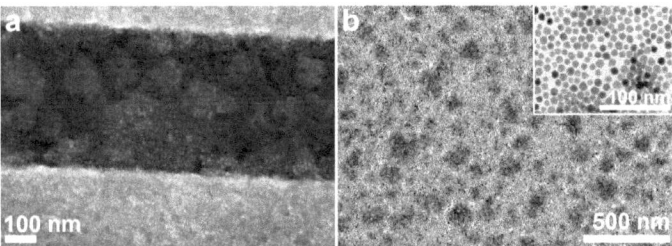

Figure 4.6: (a) SEM and (b) TEM images of a CdSe nanopyramid array. The inset in (b) shows a higher magnification TEM micrograph of the film.

and can be attributed to charging effects in the sample, for instance by trapping of carriers in defect states [132, 303]. Under irradiation of the devices with a laser at 637 nm under -10 V applied voltage, the photocurrent varied linearly with the laser intensity that was modulated with time in successive on/off cycles. Such a linear dependence occurs when exciton dissociation is the rate limiting step and many excitons recombine while they are still associated (geminate recombination) [300]. The response time between the dark current and the point when the photocurrent reaches the maximum was 2.3 s. The amplification factor from dark to photocurrent at 19.4 mW/cm^2 was 55. No significant difference was observed between source and drain currents, which implies little built-up of positive charges in the film, in contrast to results with bar electrodes separated by 0.6 µm [132]. The current fell back to the level of the dark current measured before in opposition to prevailing currents recorded with three dimensional arrays after switching off the light [312, 347]. The wavelength dependency of the photocurrent was obtained under monochromator modulated irradiation (intensity: 340 µW/cm^2 for the 560 nm component). The first absorption maximum of the CdSe nanopyramids in solution lay at 665.6 nm with a slope starting at shortly above 700 nm. The photocurrent begins to rise already at 850 nm with a small maximum at 765 nm; a second, steeper rise was noted from wavelengths similar to the onset of nanoparticle absorption. The two smooth bands of the absorption spectrum fall into one big feature in the photocurrent. A slightly increased current which does not fit with the absorbance spectrum occurs from 700 to 800 nm and is a sign for trap states below the conduction band. A more detailed explanation for the origin of this feature requires further investigation outside the scope of this work. The decrease visible after 550 nm is caused by the characteristic intensity profile of the Xe-lamp monochromator set-up.

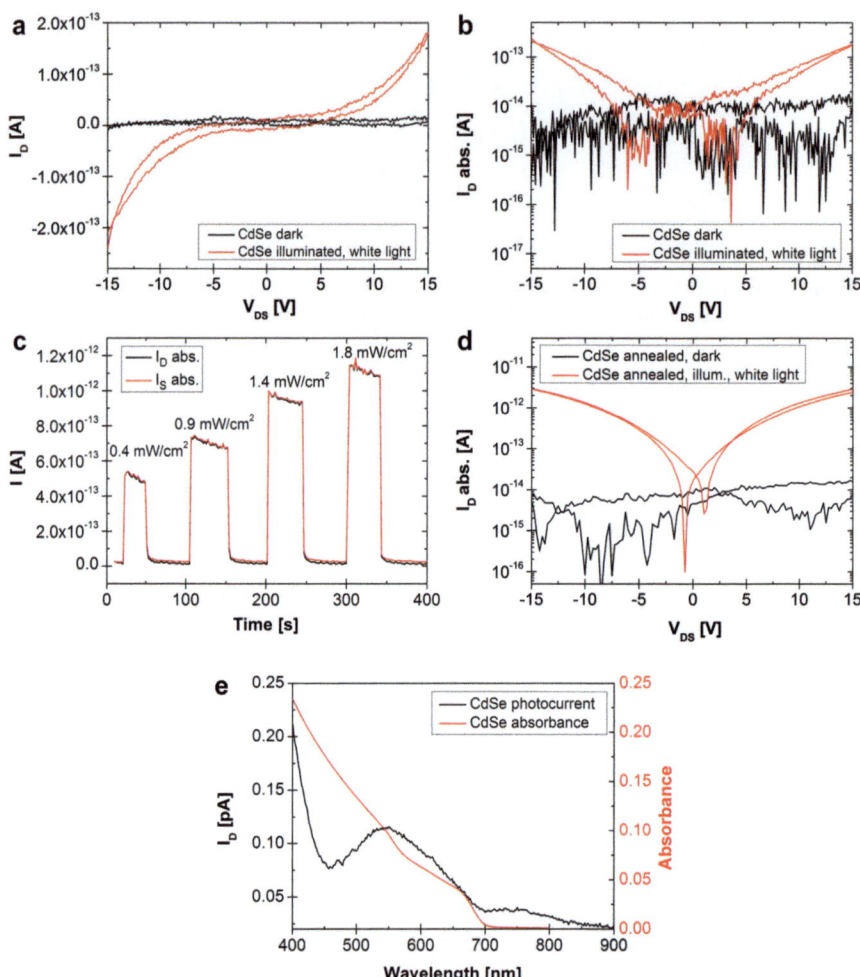

Figure 4.7: (a, b) Room temperature current-voltage characteristics of a CdSe nanopyramid device with and without white light illumination, (c) laser induced on/off photocurrent time trace at V_{DS} = -10 V and different irradiation intensities at 637 nm and (d) current-voltage curves with and without white light illumination of the same film annealed at 300 °C. (e) The wavelength dependency of the photocurrent complies with the absorbance spectrum of the pure particles in solution, apart from an additional feature at higher wavelengths.

Annealing of CdSe nanoparticle devices at 300 °C was shown to result in an optimum increase of photocurrents in two- and three-dimensional arrays by reduction of the interparticle distance and formation of preferred paths in references [132, 303]. While this process reduced the barrier widths, further heating to higher temperatures increased the conductivity through carbonisation of ligands and no photocurrent could be detected any more in two-dimensional arrays. In the devices studied here, the dark current did not increase significantly after annealing for 30 minutes under vacuum. The photocurrent on the other hand showed an expected increase of two orders of magnitude compared with the dark current, despite a deformation of the electrode structure during heating (Figure 4.7 d, a picture of the electrode can be found in the Appendix). A smaller but noticeable hysteresis remained (two minima of the absolute current that appear above and below zero bias), indicating prevailing charge trapping.

In total, the prepared CdSe devices behaved similar to the ones with bar electrodes examined before, save that with the shorter distance between the interdigitated electrodes a reduced built-up of charge under photocurrent generation was observed.

4.2.2 Electrical transport through Pt-CdSe hybrid nanoparticles

Monolayer devices of Pt-CdSe hybrid nanoparticles with two different Pt domain sizes, 1.7 nm and 3.2 nm, were examined. The nanoparticles were prepared following the procedure described in Chapter 3. Figure 4.8 shows TEM images of employed particles and SEM micrographs of the assembled films on electrodes. The compressed film with larger Pt domains exhibits a few holes but is otherwise dense, while the film with smaller Pt domains rather consists of patches connected by paths. The larger nanoparticles are, the more difficult is their assembly [348]. With metallic domains on top the rearrangement might be even more hindered. In addition to this, the combination of the ligand oleylamine and toluene might lead to lower ordering due to ligand desorption during the assembly on the subphase [346]. Particles purified less formed a dense layer after self assembling on diethylene glycol. A drawback with this layer was that a few Pt nanoparticles, formed as side products in the reaction, could be seen. Further purification otherwise lead to their removal. A treatment with additional free ligands would be counter-productive as it would increase the amount of insulating organic compounds on the device. An optimisation of the purification procedure or an exchange of the organic solvent and ligands on the nanoparticles could improve the assembly; bifunctional ligands might even connect the particles and reduce the interparticle distance. In case of the smaller Pt domains the concentration

CHAPTER 4. ELECTRICAL TRANSPORT IN HYBRID NANOPARTICLE FILMS

of the deposited particles was not high enough to form a dense layer. Nevertheless, there was a number of paths between the electrodes. The results obtained with this array will be shown since the qualitative results contribute to the understanding of the system.

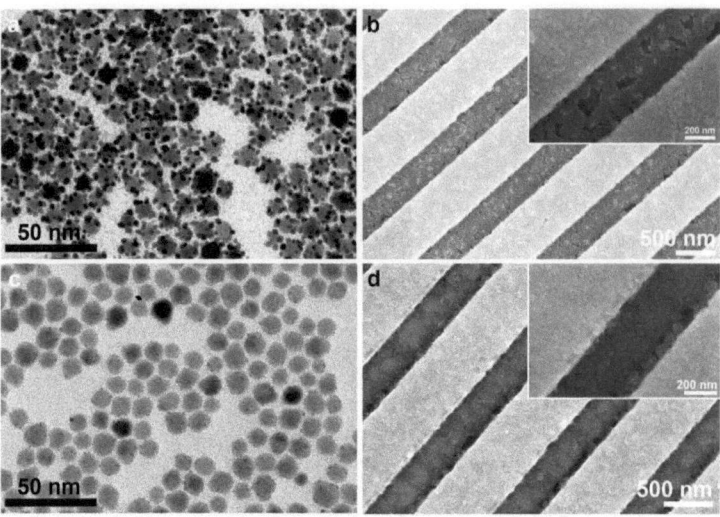

Figure 4.8: (a, c) TEM micrographs of the employed Pt-CdSe hybrid nanoparticles and (b, d) SEM micrographs of examined nanoparticle arrays on electrode structures.

4.2.2.1 Electrical transport in Pt-CdSe hybrid nanoparticles with 1.7 nm Pt domains

In areas of close contact the center-to-center distance between Pt-CdSe nanoparticles with 1.7 nm sized Pt domains was 11.6 ± 1.4 nm. The low coverage and small amount of paths manifested itself in a low dark current close to the noise level. The sample did respond to illumination with white light in the field dependent transport region with applied voltages of above 10 V (Figure 4.9). When irradiated with a laser at 637 nm and an applied voltage of 10 V, photocurrents in the regime of 10^{-13} A with an amplification factor of 32 in relation to the level of dark current and a response time of 2.3 s were recorded. With higher laser intensity the photocurrent was furthermore increased at lower voltages. With these findings the hybrid structures with small Pt domains showed a high voltage dependency presumably due to the large disorder in the film but exhibited a photocurrent generation similar to pure CdSe.

4.2 Electrical transport in CdSe and Pt-CdSe nanoparticle devices

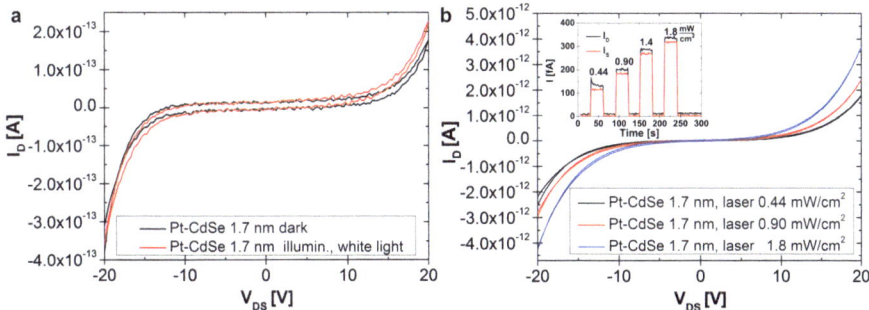

Figure 4.9: (a) Dark and photocurrents of Pt-CdSe nanoparticles with 1.7 nm sized Pt domains. (b) Irradiation intensity dependent photocurrent. Inset: $V_{DS} = 10$ V.

4.2.2.2 Electrical transport in Pt-CdSe hybrid nanoparticles with 3.2 nm Pt domains

For arrays with larger Pt domains the picture is different. Current voltage curves of films containing CdSe nanopyramids with 3.2 nm domains and center-to-center distances of 12.8 ± 1.6 nm showed remarkably different characteristics than CdSe films. In contrast to the data of pure CdSe samples, the current-voltage curves obtained in darkness already lie in the range of nA to µA. At an applied voltage of $+1$ V the dark currents of different contacts and devices ranged from 2.6 to 9.2 nA. Even though the devices behave non-ohmically, the curves do not show a hysteresis. An absence of the latter indicates that despite the high number of interfaces and potential trap states in the nanoparticle array no significant charging effects occur during the curse of current-voltage measurements. The higher dark currents are signs for improved charge injection into the film, most likely due to a bridging effect of Pt between Au and CdSe at the electrode-film contact. In bulk contacts, the work function of Pt lies below the one of Au, a circumstance that favours electron flow from the electrode onto Pt domains and thus into the nanoparticle array (see Chapter 3, Figure 3.1). A comparison between four-point and two-point probe measurements with the employed electrodes and lock-in technique (for a sketch see Appendix A) revealed that the contact resistance between Au electrodes and the nanoparticle film is negligible, so that the conductivity of the device is controlled by the resistance of the film. Wavelength resolved irradiation yielded a photocurrent without distinct features. In order to observe a significant photocurrent laser irradiation was necessary. At a constantly applied voltage of -10 V a dark current of around 4.7×10^{-7} A was recorded, which decreased by 0.1×10^{-7} A over a period of 3750 s. Such a reduction of current with time may occur due to charging processes that lead to an increased resistance in the film because of Coulomb interactions [315]. Same as with pure CdSe, a linear enhancement of photoresponse with intensity occurred, whereby a maximum amplification from dark to photocurrent of 1.016 was reached at $19.4\,\mathrm{mW/cm^2}$. The linear increase again identifies the exciton dissociation as the critical factor for the photocurrent. Even though the latter was small in comparison to the dark current, it still ranged in the order of nA, which is a remarkable gain compared to pure CdSe films.

After annealing at 300 °C the Pt-CdSe nanoparticles exhibited enhanced dark currents of up to 2.5 orders of magnitude increase compared with an untreated array. Possible reasons for this are the carbonisation of ligands, slight shifting of the nanoparticles and welding of Pt domains between adjacent nanoparticles. Signs for these processes were seen in the *in situ* annealing TEM study, discussed in Chapter 3. For comparison, until up to 400 °C annealing temperature of cobalt platinum nanoparticle arrays carbonisation of the ligands was found to be the major factor for a gain of conductivity, while it was observed to occur spontaneously above 400 °C with CdSe nanopyramids [132, 289]. The improvement of conductivity with annealing temperatures up to 300 °C in the latter case was ascribed to the formation of preferred paths and thus reduced barrier widths between the nanoparticles. The presence of metallic domains and a catalysed carbonisation of ligands at temperatures where the nanoparticles do not shift in their position may help to improve conductivity in hybrid nanoparticle films. This would remove the need for additional post synthetic treatments and may even protect the nanoparticles from environmental influences.

Taking a closer look at possible mechanisms, the superlinear increase of the room temperature current with the applied voltage can be explained by barrier height lowering. In hybrid and pure metal nanoparticles similar to the present system thermionic emission, nearest neighbour hopping (Neugebauer-Webb)/Frenkel-Poole emission and variable range hopping (Abeles) models were applied to explain the measured properties [13, 158, 348]. Fitting the current data according to these theories points to two complementary models. Barrier lowering at moderate bias (until around 2 V) may be explained by the impact of image forces on thermionic emission at Schottky barriers following Sheldon *et al.* [13]. Barrier lowering due to the electric field as in Frenkel-Poole emission is commonly reported in nanoparticle arrays under fields of more than $10^{-7}\,\mathrm{V/m}$ [331], corresponding to voltages of 5 V and above in the examined devices. An overlap of the two is likely to occur at voltages in between.

4.2 Electrical transport in CdSe and Pt-CdSe nanoparticle devices

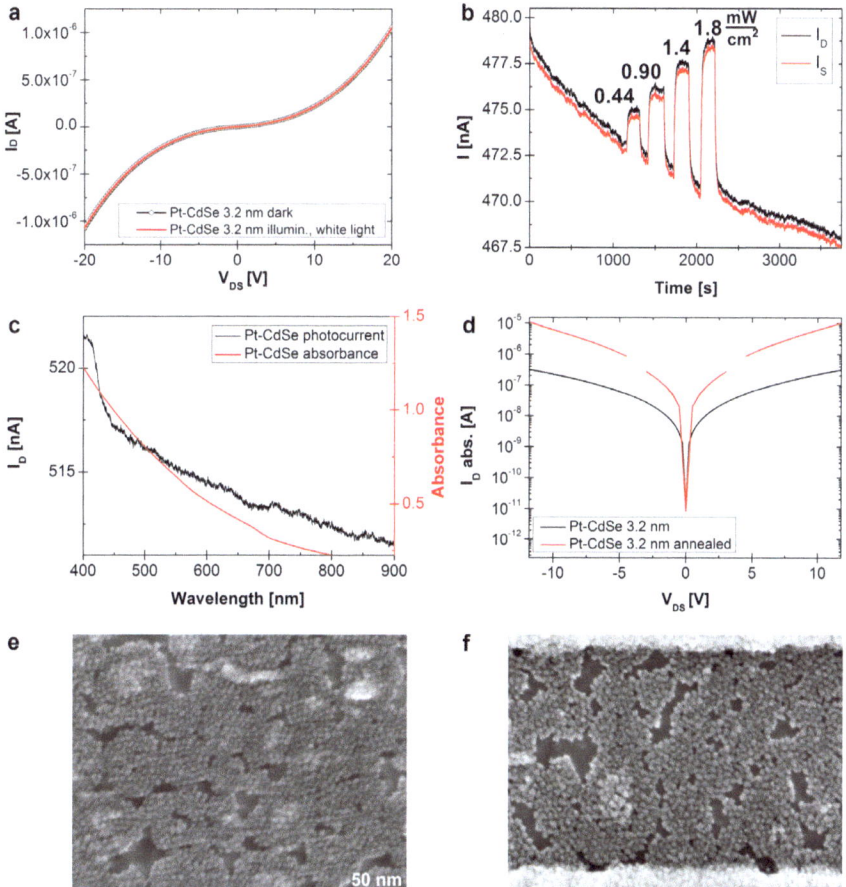

Figure 4.10: (a) Room temperature current-voltage characteristics of a CdSe nanopyramid array with 3.2 nm sized Pt domains with and without white light illumination. (b) Laser induced on/off photocurrent time trace at $V_{DS} = -10$ V and different irradiation intensities at 637 nm. (c) No defined features occur in the wavelength resolved photocurrent, plotted with the absorbance of Pt-CdSe nanoparticles in solution. (d) Absolute dark currents of the film before and after annealing at 300 °C and SEM micrographs of sections of devices before and after annealing. The gaps in the curve after annealing are due to a change in the detection range of the detector.

CHAPTER 4. ELECTRICAL TRANSPORT IN HYBRID NANOPARTICLE FILMS

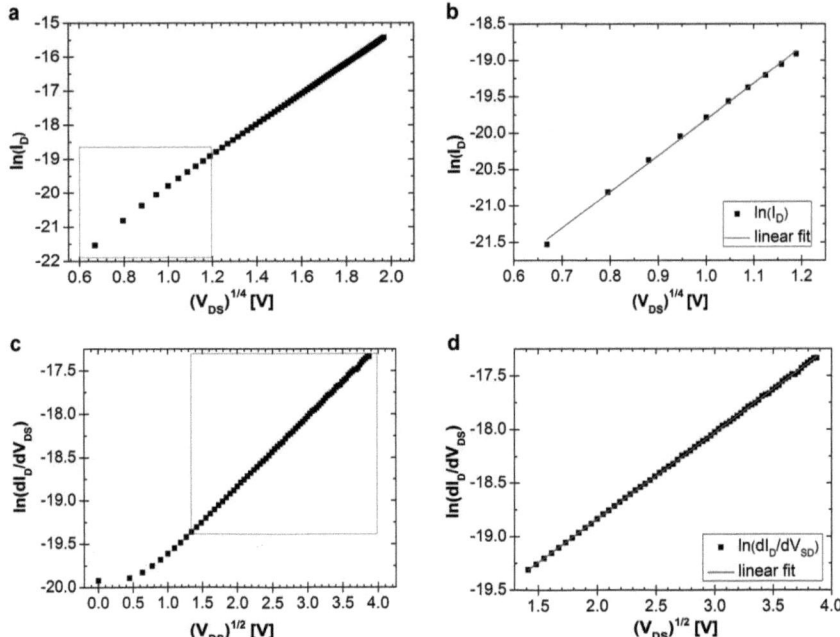

Figure 4.11: (a) Plot of the voltage dependence of the natural logarithm of the current after the thermionic emission model under consideration of a voltage dependent image force. (b) Enlarged area marked in (a) with linear fit (0.5 to 2 V). (c) Voltage dependence of the differential conductance in a Frenkel-Poole plot and (d) enlargement of the higher voltage region (2 to 15 V) with linear fit.

4.2 Electrical transport in CdSe and Pt-CdSe nanoparticle devices

To demonstrate a voltage dependence in the frame of the Schottky-Richardson theory, an image force is integrated into Equation (4.6) [13, 298]. At room temperature, where Φ_{BE} and A^* are temperature independent, the effective Schottky barrier is

$$\Phi_{BE} = \Phi_0 - \sqrt{\frac{qE}{4\pi\varepsilon_S}} \qquad (4.7)$$

where the ideal barrier height Φ_0 is reduced by the image force which contains the semiconductor permittivity ε_S. The image force in turn depends on the maximum electric field E at the junction with

$$E = \sqrt{\frac{2qN_D}{\varepsilon_S}\left(V + \Phi_{bi} - \frac{k_B T}{q}\right)} \qquad (4.8)$$

where N_D is the doping concentration in CdSe and Φ_{bi} the built-in potential. The latter is determined by the difference of the conduction band energy at the interface and in its unbent state inside the semiconductor. If it is smaller than the applied voltage, there is a linear relation between the natural logarithm of the current $\ln(I_D)$ and the applied voltage as $V^{1/4}$.

In Figure 4.11 a and b the room temperature data of a non-annealed sample plotted in this way is shown with an enlargement of the region between 0.5 to 2 V, where a linear fit reaches high compliance. At higher voltages, the slope of the data is different which suggests another mechanism. Above 2 V, the natural logarithm of the room temperature differential conductance scales linearly with $V^{1/2}$, as shown in the corresponding plots in Figure 4.11 c and d. This linear relation indicates thermally activated and field controlled Frenkel-Poole charge emission from trap states or thermally activated nearest neighbour hopping, as observed for cobalt platinum monolayers at comparable fields [348].

To gain more information about sample characteristics and possible transport mechanisms, current-voltage curves were obtained at different temperatures down to 78 K. The corresponding differential conductance curves, the conductance at different voltages plotted against temperature and fits according to different temperature dependencies are shown in Figure 4.12. Towards lower temperatures the differential conductance curves flatten around zero applied voltage, even though there is still a conductance in the regime of 10^{-11} A at 78 K. Plotting the thermal dependence of the natural logarithm of the conductance at different applied voltages reveals two facts: first, the temperature coefficient is positive over the whole range and second, the temperature dependence is affected by the applied voltages and becomes more linear with increasing voltages. From this, one might expect a

CHAPTER 4. ELECTRICAL TRANSPORT IN HYBRID NANOPARTICLE FILMS

trend towards Coulomb blockade processes at lower temperatures but over the whole examined temperature range thermally activated processes play a role. In Figure 4.12, plots created according to three different transport mechanisms are shown. The temperature dependence of thermionic emission, Arrhenius-like processes and variable range hopping follows from rearrangements of Equations (4.6), (4.4) and (4.5) and is listed in Table 4.1.

Table 4.1: Temperature dependencies of selected transport mechanisms.

Mechanism	Temperature dependence
Thermionic emission	$\frac{ln(I)}{T^2} \propto \frac{1}{T}$
Arrhenius-like/NNH	$ln\left(\frac{dI}{dV}\right) \propto \frac{1}{T}$
VRH	$ln\left(\frac{dI}{dV}\right) \propto T^{(1/(D+1))}$

The curve shapes of the temperature dependence at zero and 1 V applied voltage are very similar. Since the thermionic emission plot cannot be evaluated with the values obtained with no applied voltage, the temperature dependent data at 1 V applied voltage was employed for analysis. The best fit in all temperature regions was obtained for the variable-range hopping mechanism with a dimension $D=1$. A fit with $D=2$ deviated significantly and is not shown. Nevertheless, the thermionic emission fit does not deviate strongly and the barrier height and the Richardson constant are only constants at around room temperature [298]. The fit in the region between 150 K to 290 K hints towards a linearity but no further assumptions can be made owing to the small number of data points.

Discarding Arrhenius-like behaviour associated with nearest neighbour hopping might be too early, since the fit would converge better without the value at 78 K (12.9×1000 T^{-1}) and it is possible that different processes occur in consecutive temperature regions. To clarify this and to make secure statements, further studies with higher resolution below and also above the temperature range examined here are necessary and pursued. Such data would furthermore allow for the calculation of reliable barrier heights or activation energies, which in turn should help to rationalise the occurring mechanisms.

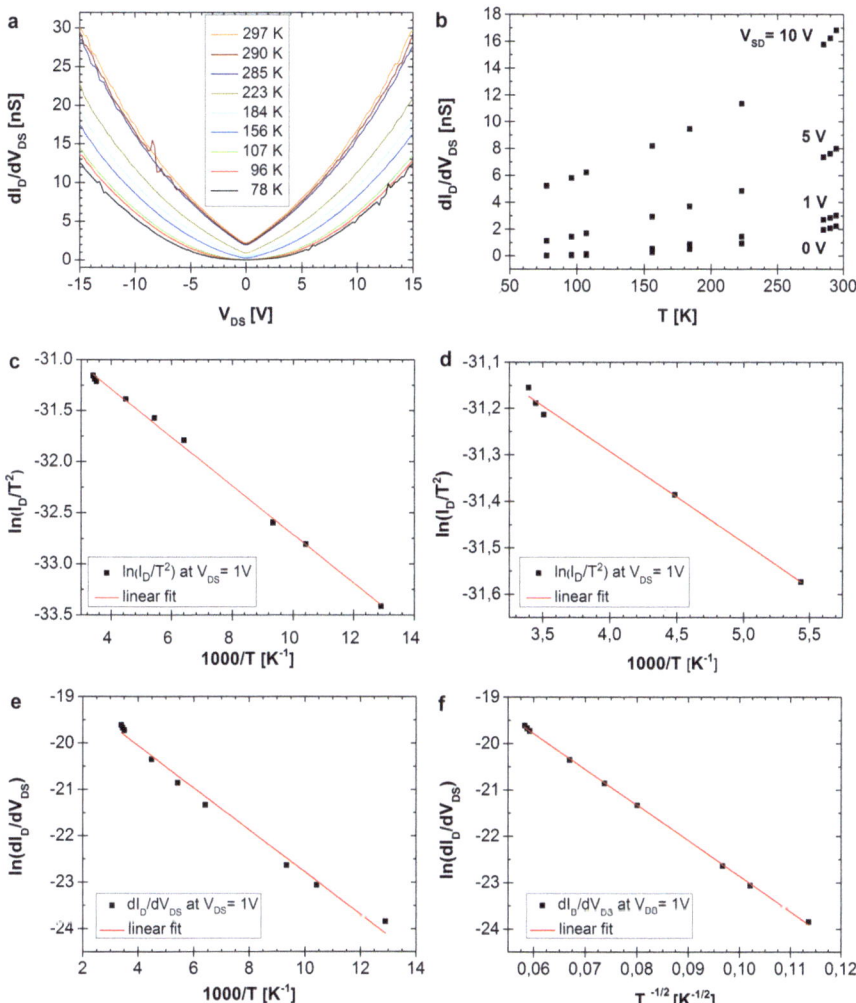

Figure 4.12: (a) Plots of the differential conductance at varied temperatures and (b) the voltage dependent relation of conductance and temperature. (c) The temperature dependence of the current at 1 V plotted after the thermionic emission theory and (d) the enlarged region from 150 to 290 K. (e) The differential conductance at 1 V plotted for a simple Arrhenius-like activated hopping and (f) a variable range hopping process with D = 1.

CHAPTER 4. ELECTRICAL TRANSPORT IN HYBRID NANOPARTICLE FILMS

To understand the results, one has to think of the possible paths a charge carrier could take inside the array. In principle, two extreme scenarios of transport through the array are possible: one is charge movement across the array with alternate stays on both, the metal and the semiconductor component, and barriers at the interfaces. The other scenario would be transport *via* transitions between the metallic domains, dominated by Coulomb interactions. Figure 4.13 depicts both these paths, whereby different possible transitions involving the semiconductor component are shown for the first option.

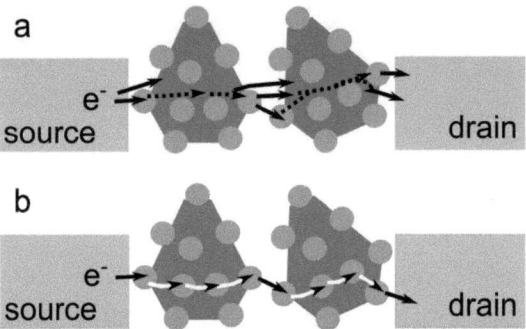

Figure 4.13: Possible paths for electron transfer, (a) with metal to semiconductor transitions and (b) solely between metal domains.

Irrespective of the path, barriers between the nanoparticles have to be overcome by tunnelling or hopping. It is likely that Pt is taking part at least on one side of the barrier because of its exposed position on the nanoparticle and proximity to the adjacent particles. The existence of a high dark current in arrays with large Pt domains in contrast to pure CdSe arrays points to a Pt dominated transport. The photocurrent generated in CdSe would then be a simple addition to the dark current. Compared with photocurrents in pure CdSe arrays it is orders of magnitude higher, which might be explained by the generally higher conductance and also an improved charge separation due to the metal-semiconductor interfaces. Another contribution could be light scattering by the metal [345].

Unknown elements in these considerations are the relative positions of the electronic levels in the two nanoparticle components and the role of possible metal induced gap states. As mentioned in Chapter 3, the reported values for band edge positions of CdSe bulk and nanoparticles vary strongly and the work function of the metal depends on the exposed surface. Thus it is difficult to predict relative band levels of CdSe and Pt nanodomains. Based on the bulk values, the work function of Pt should be positioned inside the band gap of CdSe slightly closer to the conduction band but with an energetic difference in the regime of hundreds of meV. This would create a substantial driving force for excited electrons to be transported away from the semiconductor. Regarding the dark current, such an energetic difference would have to be overcome by thermal activation or tunnelling. The prior would explain the signs for voltage dependent thermionic emission at lower fields. The reason for this is that the limiting barriers must be metal semiconductor interfaces inside the hybrid nanoparticles because the resistance of the electrode-film contact was proven to be negligible. The compliance with the temperature dependence of variable-range hopping and Frenkel-Poole emission at high fields might be interpreted as Coulomb force controlled charge transfer processes between Pt domains. This may even occur across CdSe to the next Pt domain with the semiconductor as a lower barrier compared to vacuum or ligands.

4.3 Conclusions and perspective

Monolayers of colloidal CdSe semiconductor and Pt-CdSe metal-semiconductor systems on interdigitated electrodes were prepared and examined in terms of their electrical transport properties. The results with CdSe nanoarrays confirmed known characteristics such as wavelength and voltage dependent photoconductance which is tunable through the intensity of the incoming irradiation.

The assembly of monolayers from hybrid nanoparticles has proven to be a critical point for the device fabrication and characterisation. Increasing the homogeneity of the films in future investigations might be reached, for example, by modifications in the purification procedures, in the solvent used for deposition on the subphase or even ligand exchange on the nanoparticles to shorter or bifunctional molecules.

In contrast to pure CdSe nanoparticles, Pt-CdSe hybrid arrays with 3.2 nm sized Pt domains exhibited a significant dark current in the regime of 47 µA at an applied voltage of 10 V which could be further increased by annealing. Under laser irradiation an intensity dependent photocurrent was generated on top of the dark current. Room temperature current-voltage curves were affected by barrier lowering processes. A combination of thermionic emission related to barriers at metal-semiconductor interfaces at low voltages and thermally activated Frenkel-Poole emission from localized states at high voltages with a dominating role of Pt is suggested as an explanation. Temperature dependent data obtained from 78 K to 300 K points to hopping processes in the charge transport at reduced temperatures. Such strong influence was not observed in Pt-CdSe nanoparticle arrays with the smaller 1.7 nm Pt domains hinting at a tunability of Pt influence by its size. Anyhow, the low density of the nanoparticle film lead to such small currents that this result should be treated with care. By studies with different sizes of Pt domains and possibly other metals the presumption of varying dominance of Pt in transport processes may be tested.

In order to reliably distinguish between transport processes and their temperature dependent occurrence, investigations with higher resolution of measurements at varying temperatures from below 78 K up to 325 or 375 K should be conducted. Possible carbonisation processes during annealing to higher temperatures may furthermore be taken advantage of in order to increase the conductance without chemical post-synthetic treatments.

Additionally, it might be clarified if effects such as Fermi level pinning and Coulomb blockades play a role in the system. For this, the above mentioned studies could be complemented by investigations of gate dependent (transistor-like) behaviour of the samples to identify leakage currents caused by metal induced gaps states. Furthermore, scanning tunnelling microscopy may provide insights into the electronic structure of the hybrid nanoparticles. This will not only support the interpretation of the electrical transport but also contribute to the understanding of interactions in the little examined electronic ground state of oligomeric hybrid nanoparticles.

Successive variations of the particle components and compositions with respect to their theoretical band alignment and the metal domain size will be an opportunity to transfer the obtained results and possibly lead to nanoparticle arrays with improved properties for optoelectronic applications.

5 Experimental

5.1 Materials and preparation methods

5.1.1 Materials

Cadmium oxide (CdO; 99.99+%) was bought from ChemPur. Tri-n-octylphosphane (TOP 90% pure and vacuum distilled, stored in a nitrogen filled glovebox), 1-chlorooctadecane (COD; 96%), 1,2-dichlorobutane (1,2-DCB; 98%), 1,1,2-trichloroethane (TCE; 96%), 1,2-diiodoethane (DIE; 99%), n-dodecyltrimethylammonium chloride (DTAC; >99%), gold(III) chloride (AuCl$_3$; 99%, stored in a nitrogen filled glovebox), oleylamine (*cis*-9-octadecenylamine, OAm; 70% tech., filtrated through a 0.45 μm PTFE syringe filter by Carl Roth prior to use), palladium(II) acetate (Pd(ac)$_2$; ≥99.9%), tetra-n-butylammonium borohydride (TBAB; 98%), hexadecylamine (HDA, 90%), dodecylamine (DDA; 99.5%) and selenium shots (amorphous, 2-4 mm, 0.08-0.16 in, 99.999+%, stored in a nitrogen filled glovebox) were obtained from Sigma-Aldrich. Tri-n-octylphosphane oxide (TOPO; >98%), 1,2-dichloroethane (DCE; p.A.), Acilit pH paper, acetone (p.A.), methanol (p.A.), ethanol (p.A.), 2-propanol (p.A.), trichloromethane (p.A.), n-hexane (p.A.) and toluene (p.A.) were aquired from Merck. Acros is the producer of 1-dodecanethiol (DDT; 98%), 1,2-dibromoethane (DBE; 99%), oleylamine (OA; 80-90% C18 content, stored under nitrogen atmosphere) and 2,3-dichlorobutane (2,3-DCB; 98%), while n-dodecyltrimethylammonium bromide (DTAB; 99%, vacuum dried and stored in a nitrogen filled glovebox), oleic acid (OAc; 99%) and n-octadecylphosphonic acid (ODPA, 93 and 97-98%) stem from Alfa Aesar. Further batches of ODPA were bought from PCI (ODPA; ≥99%). Platinum(II) acetylacetonate (Pt(acac)$_2$; 98%) and palladium(II) acetylacetonate (Pd(acac)$_2$; 99%) were bought from abcr. Silver(I) acetate (Ag(ac)$_2$; 99 %) was obtained from Strem. Highly oriented pyrolytic graphite (HOPG) substrates of ZYB quality ($10 \times 10 \times 2\,\text{mm}^3$) were purchased from NT-MDT. The chemicals were used without further purification if not indicated else. Teflon tape was procured from VWR.

5.1.2 Synthesis of pyramidal CdSe nanocrystals

5.1.2.1 Standard recipe

The reactions were carried out under nitrogen and application of standard Schlenk techniques. The temperatures of injection and growth as well as reaction times were varied, the changes are notified in the relevant section. The following standard protocol was derived from [16] and reported in [105] with minor modifications. A 1 M selenium stock solution was prepared inside the glove box by dissolving 1.580 g (20.0 mmol) Se in 20 mL of tri-n-octylphosphane (TOP) under stirring at room temperature over night. For the nanoparticle synthesis, a three necked flask was equipped with a condenser, a quickfit with septum and a glass covered thermocouple. Inside, 25 mg (0.19 mmol) CdO, 0.14 g (0.42 mmol) n-octadecylphosponic acid (ODPA) and 3.0 g (7.8 mmol) tri-n-octylphosphane oxide (TOPO) were heated to 120 °C for 30 min *in vacuo*, during which two switches to nitrogen were carried out. At 270-290 °C Cd was complexed until the solution was clear and colourless (around 60 min). At 77-80 °C, 10 µL (0.13 mmol) of 1,2-dichloroethane (DCE) were injected carefully with a 50 µL Hamilton syringe with extra long needle (11 cm) before the temperature was raised again. At 265 °C, 0.42 mL of Se in TOP (1 M, 0.42 mmol Se) were injected. For growth the temperature was set to 255 ± 2 °C. Nanoparticle formation was visible by colour transitions from colourless over yellow and orange-red to brown. After 4 h, the reaction was quenched by cooling down to 75 °C and injecting 3.5 mL of toluene. Synthesised CdSe nanopyramids were transferred into airtight vials in portions of 1 to 2.5 mL, sealed with teflon tape and stored in a freezer at -8 °C. For characterisation, aliquots of the product and samples taken during the reaction were washed by three cycles of precipitation/re-dispersion with toluene/methanol 2:1 and centrifugation (3 min at 4500 rpm/1856 g). Intermediate supernatants were discarded. After the final cycle the samples were re-dispersed in toluene. *Note:* Different batches of ODPA were tested with the result that old or less pure ones inhibited the shape evolution, possibly due to competition of impurities with chloride. To avoid deviations caused by contaminants it is recommended to use the highest quality of all reagents and distilled TOP.

5.1.2.2 Variation of the halogenated additives

The experiments were based on the procedure in section 5.1.2.1. Halogen sources other than DCE were added in a ratio to Cd of 0.7 at 10 °C below their boiling point, at 80 °C in case of n-dodecyltrimethylammonium chloride (DTAC) or before the injection of Se in

TOP at 265 °C when 1-chlorooctadecane (COD) was employed. For details see Table 5.1. *Warning on the use of 1,2-diiodoethane (DIE) in the reaction:* The strong acidity of HI that may form in the reaction with DIE possibly leads to the production of H$_2$Se. The gas may be carried out of the apparatus with the inert gas flow when conducted in a laboratory hood.

Table 5.1: Volumes V_{add}, masses m_{add}, molar amounts n_{add} and injection temperatures T_{add} for different halogenated additives.

Additive	V_{add} [µL]	m_{add} [mg]	n_{add} [µmol]	T_{add} [°C]
1,2-Dibromoethane	11.0	23.8	127	122
1,2-Diiodoethane	-	35.5	126	190
1-Chlorooctadecane	-	32.5	123	80
1,1,2-Trichloroethane	11.5	16.6	124	103
1,2-Dichlorobutane	14.5	16.1	127	106
2,3-Dichlorobutane	14.5	16.1	127	107
n-Dodecyltrimethylammonium chloride	42.5	36.1	125	265

5.1.2.3 Injection of 1-chlorooctadecane after CdSe nucleation

The reaction was carried out similarly to the method described in subsection 5.1.2.1, save that COD was added at the growth temperature of 255 °C 15 min after the injection of Se in TOP at 300 °C. The growth phase was maintained for 24 h during which several aliquots were taken.

5.1.2.4 Determination of the relative concentration of protons in aliquots

At different stages of the reaction, aliquots of 0.15 mL were sampled and diluted with 0.5 mL of toluene. Of this dispersion, 300 µL were vortexed with 150 µL of distilled water, shaken for further 15 min and left to separate for 1 h. Afterwards, the samples were centrifuged for 2 min at 1000 rpm. Drops of the aqueous layer were deposited onto wetted Acilit pH paper to determine their pH value.

5.1.2.5 Samples for XPS

The set-up was the same as in section 5.1.2.1, the method follows reference [125]. Differently to 5.1.2.1, 0.20 g (0.42 mmol) ODPA were used and a freshly peeled piece of highly

CHAPTER 5. EXPERIMENTAL

oriented pyrolytic graphite (HOPG) was inserted with the solid compounds before complexation. Once the solution was colourless (after around 60 min), 3.0 µL (42 µmol) DCE or 3.3 µL (42 µmol) of DBE were injected at 80/127 °C. The period of growth lasted for 22 h. The reaction was quenched by cooling down to 70 °C and injecting (3.0 mL) of toluene. The nanoparticle dispersion was removed with a syringe and the HOPG substrates were freed from excess solvents and nanoparticles by careful rinsing with toluene and dipping into three fresh baths of toluene. Afterwards the samples were dried and stored under nitrogen. Nanoparticles from the supernatant were purified for characterisation as described in subsection 5.1.2.1.

5.1.3 Synthesis of Au-CdSe pyramid hybrid nanoparticles

The deposition of gold was carried out under nitrogen and ambient conditions following recipes derived from the procedures reported in [17] and [105].

5.1.3.1 Gold(III)-stock solutions

Gold stock solutions were prepared by dissolving gold(III)chloride and n-dodecyltrimethylammonium bromide (DTAB) (1:1.6, weighed out inside the glovebox) in toluene under nitrogen atmosphere. After mild sonication (2 min, 50%) and slight warming (heating plate, 50 °C) a clear, orange-red solution was obtained. Typically, 20.1 mg $AuCl_3$ (66.3 µmol) and 32.0 mg (104 µmol) DTAB were dissolved in 15.0 mL of toluene to give a 4.46 mM solution with respect to Au(III). The resulting stock solution was stored in darkness.

5.1.3.2 CdSe dispersions

CdSe nanopyramids were taken out of the freezer and left to warm to room temperature before they were purified by precipitation and re-dispersion cycles. For this, 1 mL of CdSe dispersion was precipitated with 0.5 mL methanol and centrifuged at 4500 rpm for 3 min. After the colourless supernatant was discarded, the nanoparticles were re-dispersed in 1 mL of toluene as a stock dispersion for several experiments. With CdSe nanopyramids prepared under addition of COD, this process was repeated two more times, before the purified particles were dispersed in 2 mL of toluene. Of the purified dispersion, 100 µL were diluted to 3 mL with toluene in a quartz cuvette and the optical density at the first absorption maximum was determined (see section 5.2.2). Multiplied by 30, this value gave the optical density of the dispersion OD_{CdSe}.

5.1.3.3 Au-CdSe hybdrid nanoparticles with DCE-nanopyramids

The reactions were carried out under nitrogen and ambient atmosphere at room temperature (22 °C). No morphological difference was observed between the products prepared under different atmospheres. For a standard Au deposition on CdSe nanopyramids prepared with DCE, a volume of purified CdSe dispersion ($V_{CdSedisp.}$) was diluted to 4 mL with toluene to achieve an optical density of 0.27 (($V_{CdSedisp.}*OD_{CdSe}$)/4 mL) [105]. Gold precursor solutions with different molar amounts n_{Au} of Au were prepared by diluting volumes of the stock solution with toluene to 4 mL and mixing with oleylamine (OAm) or dodecanethiol (DDT) with a ligand to Au ratio of 23. The Au/CdSe ratio was calculated from the micromolar amount of Au divided by ($V_{CdSedisp.}*OD_{CdSedisp.}$) and reactions were carried out with 0.96, 1.60, 1.92, 2.88 and 3.84. To obtain a ratio of 1.92 with DDT as ligand, for example, 393 µL of a 1.6 M Au(III)-stock solution (2.07 µmol) were diluted with 3.61 mL of toluene and mixed with 10 µL (48.9 µmol) of DDT. For the same ratio but with OAm (70%), 18.7 µL (48.9 µmol) of the ligand were necessary. After 5 min of sonication the gold precursor was injected swiftly into the CdSe receiver under fast stirring. The reactions were left to run for 1 h prior to three cycles of precipitation with ethanol (1:1), centrifugation (3 min at 4500 rpm) and re-dispersion in toluene for further characterisation. For an additional reduction of deposited Au before the purification, sufficient volumes of a freshly prepared 4.0 mM solution of tetra-n-butylammonium borohydride (TBAB) and DTAB in toluene (1:1; 10.1 mg TBAB and 12.1 mg DTAB in 10 mL of toluene) were added to obtain an Au/TBAB ratio of 16 (Au to H$^-$: 4). After 20 min of violent stirring the samples were purified as described.

Stepwise injection: The growth and ripening of Au domains with increasing Au-DDT precursor addition was examined by stepwise injection of a solution with a total volume of 4 mL containing 12.5 µmol of Au (Au/CdSe 7.68). Samples of 0.2 mL were taken 5 min after the injections and quenched with 1 mL of ethanol before a work up with toluene/ethanol (see paragraph before).

Incubation with Au(III)-stock solution under UV-irradiation: In a cuvette equipped with a magnetic stirrer, 3 mL of a CdSe dispersion with an optical density of 0.027 were irradiated with a 366 nm UV-lamp (6 mW) positioned at a distance of 3 cm. After 30 seconds 51.9 µL of a 1.36 mM Au(III)-stock solution (233 nmol) were injected under continuing irradiation. The dispersion turned turpid during 5 min of rapid stirring, indicating an agglomeration of the nanoparticles. After 10 min the irradiation was ended and the reaction solution was purified as described above.

CHAPTER 5. EXPERIMENTAL

5.1.3.4 Au-CdSe hybdrid nanoparticles with COD-nanopyramids

Nanopyramids with first absorption maxima between 667 and 671 nm were prepared by following the recipe with COD described in subsection 5.1.2.2, save that the growth was maintained for 24 h. They were purified with three cycles of precipitation and re-dispersion with toluene/methanol 1:0.5 and 3 min of centrifugation at 4500 rpm.

Method 1: "Diluted"

The reactions were carried out under ambient conditions at room temperature (22 °C). The procedure was similar to the one described in subsection 5.1.3.3. Appropriate volumes of purified CdSe nanopyramid dispersions ($V_{CdSedisp.}$) were diluted to a volume of 2 or 4 mL with toluene and an optical density of 0.27. Gold precursor solutions with different molar amounts of Au were prepared by diluting volumes of the Au(III)-stock solution with toluene to 1 or 2 mL. For an incubation with Au(I)-precursor, the latter was mixed with DDT (Au-DDT solution). Analogous to the above procedure, the ratio of Au/CdSe was calculated from the molar amount n_{Au} of Au divided by ($V_{CdSedisp.} * OD_{CdSedisp.}$). After 5 min of shaking they were swiftly injected into the CdSe receiver under fast stirring. The Au/CdSe ratios examined were 1.30, 1.96, 2.60, 2.90 and 3.90. DDT/Au ratios in Au(I)-precursor solutions were varied from approximately 10 to 23. While no large differences in terms of Au domain sizes were observed, the colloidal stability was found to be the highest with a ratio of 18. Above this value precipitation of needle shaped organic crystals was occasionally observed, especially at high Au precursor concentrations.

The reactions were left to stir for 1 h with Au-DDT and 10 min with Au(III). The shorter reaction time in the latter case resulted from the precipitation of the nanoparticles. They were then re-dispersed by adding DDT in a minimum 18-fold excess to Au. The produced hybrid nanoparticles were worked-up with three cycles of precipitation with ethanol (1:1), centrifugation (3 min at 4500 rpm) and were finally re-dispersed in toluene for further characterisation.

Treatment with DDT before Au deposition: To examine the influence of nanopyramid passivation by DDT, the same conditions were applied as if Au precursor was injected save that only DDT was added. In particular, purified CdSe nanoparticles were diluted with toluene to give a dispersion of 6 mL with an optical density of 0.18 (this value results from the calculation based on the whole reaction volume). To this dispersion 13.2/5.1 µL of DDT (53.7/20.7 µmol) were added to simulate different ligand (precursor) concentrations.

5.1 Materials and preparation methods

The mixture was stirred for 5 min before the nanoparticles were precipitated with ethanol (1:1) and centrifuged at 7000 rpm for 3 min. Two more cycles with toluene/methanol (1:1) followed. Finally the nanoparticles were re-dispersed in 1 mL of toluene. Of the sample with 53.7 µmol of DDT 0.5 mL were dried for ATR-IR measurements. The rest was diluted to 2 mL (optical density 0.27), wildly agitated and incubated with 1 mL of Au(III)-solution containing 2.67 µmol to test the deposition (Au/CdSe 4.94). Half of the nanoparticles treated with 20.7 µmol DDT were incubated with 528 nmol of Au(III) (Au/CdSe 0.98). Both reactions were run for 10 min, in which nanoparticles of the reaction with the lower amount of Au began to precipitate after half of the time. The samples were precipitated and re-dispersed with ethanol (1:1) and two cycles with toluene/methanol (1:1) before they were taken up in toluene. Centrifugation was conducted for 3 min at 7000 rpm after each precipitation. The purification procedure lead to agglomeration of the samples, as observed in other cases of incubation with Au(III)-precursor.

Incubation with an aged Au-oleylamine solution: Gold precursor solutions with 2.12 µmol of Au and 48.5 µmol of oleylamine (C18 content 80-90%) in 1 mL of toluene were prepared and shaken for 24 h. CdSe nanopyramid receivers had a volume of 3 mL with an optical density of 0.19. They were incubated with aliquots of the gold solutions to obtain ratios of 1.83 and 3.24 (0.5 and 0.88 mL). After 10 min, the reactions were quenched through the addition of ethanol (1:1). The nanoparticles were centrifuged for 3 min at 4500 rpm. Another cycle of purification was carried out with toluene/methanol (1:1) before the samples were taken up in toluene for further characterisation.

Method 2: "Concentrated"

In each reaction 500 µL of CdSe receiver with an optical density of 1.6 were incubated with volumes V_{Au} of a 4.4 mM Au(III)-stock solution under vigorous stirring. For the incubation with Au-DDT, the Au solution was mixed with the appropriate amount of ligand and shaken for 5 to 7 min during which it turned colourless. This solution was injected into the receiver and the reaction was left to stir for 1 h. For an Au-shell formation the appropriate amount of Au(III)-stock solution was added directly. When the nanoparticles had precipitated after 10 min, DDT or OAm were added and the reaction was stirred for another 30 min. The DDT to Au ratio was 18 in all cases, with OAm it was at least 23. Exemplary values of the added volumes and amounts are listed in Table 5.2. Finally, the nanoparticles were precipitated with 1 mL of ethanol, sonicated for 10 s in case of Au(I) incubation and centrifuged at 4500 rpm for 3 min. The supernatant was discarded and the nanoparticles were re-dispersed in 0.75 mL of toluene. After addition of ethanol and methanol (1:1:0.6), the samples were centrifuged again for 3 min at 7000 rpm. A final cycle was carried out with toluene/methanol (1:3) and centrifugation for 3 min at 11000 rpm (11092 g). For further characterisation the samples were re-dispersed in 0.25 mL of toluene. With oleylamine the nanoparticles could be re-dispersed homogeneously only after the first cycle of purification.

Table 5.2: Details of gold precursor solutions (volume V_{Au} of Au(III)-stock solution, molar amount n_{Au} of Au), Au to CdSe ratio and added amount of the ligand DDT (with Au-DDT added to gold precursor solution, with Au(III) added after precipitation of the nanoparticles) for exemplary reactions.

Au(III)/Au-DDT	V_{Au} [μL]	n_{Au} [μmol]	Au/CdSe	V_{DDT} [μL]	n_{DDT} [μmol]
Au(III)	520	2.30	2.88	10	40.7
Au(III)	355	1.57	1.97	7	28.5
Au-DDT	520	2.30	2.88	10	40.7
Au-DDT	354	1.57	1.97	7	28.5
Au-DDT	253	1.12	1.40	5	20.3

5.1.4 Reactions of CdSe nanopyramids with Ag, Pd and Pt precursors

Before the reaction, CdSe nanopyramids were purified three times by precipitation and re-dispersion with toluene/methanol 1:0.5 (3 min at 4500 rpm) and re-dispersed in toluene. The optical density was determined by absorption spectroscopy (see 5.2.2). The reactions were carried out in a three necked flask with a volume of 25 or 50 mL, equipped with a condenser, a quickfit with a septum and a glass covered thermocouple under nitrogen flow. Stirring was maintained at level 3 on an RH basic 2 IKAMAG stirrer. First, 31 μmol of metal acetate or acetylacetonate were dissolved in toluene. After an optically clear solution was obtained, 1.0 mL of oleylamine (C18 content 80-90%, 2.42 mmol for 80%) was added and the solution was stirred for 5 min. Then, purified CdSe nanopyramids were injected and the temperature was set to 120 - 124 °C to achieve mild reflux of toluene. The time was started when the reflux temperature had been reached. Total reaction volumes were 12.5 or 25 mL. The volume of toluene added to dissolve the metal salt was calculated after the determination of the volume of CdSe nanopyramid dispersion necessary to obtain the desired overall optical density at the first absorption maximum (in most cases 0.1) in

5.1 Materials and preparation methods

the reaction solution. The time of reaction was varied from 10 min to 24 h and ended with the removal of the heating mantle. The effect of thermal activation was examined by control reactions conducted at room temperature. The ratio of metal to CdSe was calculated from the micromolar amount n of metal divided by $(V_{CdSesolution}*OD_{CdSe})$, analogous to subsection 5.1.3.3. Element and reaction specific details are referred in the following paragraphs.

5.1.4.1 Silver

Reactions with Ag/CdSe ratios of 24 and 12 were carried out with 5.2 mg (31 µmol) silver(I) acetate in 12.5 and 25 mL of OAm/toluene with an optical density of 0.1 in regard to CdSe (nanopyramids prepared with DCE). With proceeding time under toluene reflux the brown colour of the dispersion turned darker. Aliquots were cooled down with ice to quench the reaction. They were purified by three cycles of precipitation and re-dispersion with toluene/ethanol (1:3) and 3 min of centrifugation at 7000 rpm (4492 g) before being re-dispersed in toluene.

5.1.4.2 Palladium

Palladium acetate or palladium acetylacetonate were dissolved in toluene. In case of the acetate, the solution was warmed to 40 to 70 °C after the addition of oleylamine until all solid was dissolved in a colourless solution. CdSe nanopyramids were then injected in toluene at 50 °C. Both types of nanopyramids, prepared with DCE and COD, were employed. With the acetylacetonate, a yellow solution was obtained through stirring at room temperature for 10 min. The colour persisted after the addition of oleylamine. With time the reaction dispersion turned darker and after around 5 h the particles tended to precipitate irreversibly (in some cases even earlier). The addition of further 0.5 mL of oleylamine at different reaction times could not prevent this. At room temperature the colour of the dispersion remained the same and the nanoparticles were stable. Exemplary reaction details are listed in Tables 5.3, 5.4 and 5.5.

CHAPTER 5. EXPERIMENTAL

Table 5.3: Reaction parameters for ion exchange processes of CdSe nanopyramids prepared with DCE/CdSe 0.7 and Pd(II). Listed are the type of Pd salt, its weight m_{Pd} and molar amount n_{Pd}, the volume of the reaction dispersion $V_{reac.}$, its optical density with regard to CdSe OD_{CdSe}, the ratio of Pd to CdSe and the reaction time t.

Salt	m_{Pd} [mg]	n_{Pd} [µmol]	$V_{reac.}$	OD_{CdSe}	Pd/CdSe	t [min]	T [°C]
ac	7.0	31	25	0.092	13.5	1440	110
ac	6.9	31	25	0.097	12.7	7440	22
ac	6.9	31	12.5	0.095	26	345	110
acac	9.5	31	12.6	0.10	24	1440	110

In reactions with nanopyramids prepared with COD, the volume was 12.5 mL and the reaction temperature 110 °C, other details may be extracted from Table 5.4.

Table 5.4: Reaction parameters for ion exchange processes of CdSe nanopyramids prepared with COD/CdSe 0.7 and Pd(II). Written are the type of Pd salt, its weight m_{Pd} and molar amount n_{Pd}, the volume of the reaction dispersion $V_{reac.}$, the optical density with regard to CdSe OD_{CdSe}, the ratio of Pd to CdSe and the reaction time t.

Salt	m_{Pd} [mg]	n_{Pd} [µmol]	$V_{reac.}$	OD_{CdSe}	Pd/CdSe	t [min]
ac	7.1	31	11.5	0.16	16	300

In reactions with other amines, the total volume was 12.5 mL (including the molten amine) with an optical density of 0.1 and a Pd/CdSe ratio of 25. The reaction temperature was 110 °C. Other details are listed in Table 5.5.

Table 5.5: Reaction parameters for ion exchange processes of CdSe pyramids prepared with DCE/CdSe 0.7 and Pd(II) acetate with amines other than oleylamine. Listed are the amine (DDA: dodecylamine, HDA: hexadecylamine), its weight m_{amine} and molar amount n_{amine} and the reaction time t.

Amine	m_{amine} [mg]	n_{amine} [mmol]	t [min]
DDA	0.5631	2.977	240
DDA	0.9374	3.494	150

5.1 Materials and preparation methods

The nanoparticles were treated with three cycles of precipitation and re-dispersion with toluene/ethanol (1:3) and 3 min of centrifugation at 7000 rpm (4492 g) or 15000 rpm (20627 g). Samples from late reaction stages had precipitated already and could not be re-dispersed even by adding drops of ligands (dodecanethiol, oleylamine) and sonicating them. All samples were dispersed in toluene for characterisation.

Organic and possibly precursor based residues were visible in TEM samples, so that other attempts of cleaning were made to improve the purity. Among these was ultrasonication and slight warming (40 °C) of the precipitated samples in ethanol to dissolve potentially crystallised organic side products. Changing the alcohol to methanol or isopropanol did not significantly alter the situation.

5.1.4.3 Platinum

Platinum acetylacetonate (12.1 mg, 30.8 µmol) was dissolved in toluene to yield a clear yellow solution. To this solution, 1 mL of oleylamine (C18 content 80-90 %, 2.42 mmol for 80%) was injected. After 5 min of stirring, an appropriate amount of CdSe nanopyramids (DCE) in toluene was added to result in a reaction solution of 12.5 mL with an optical density of 0.1 regarding CdSe. The mixture was heated to reflux and left to stir for 24 h during which aliquots of 1 mL were taken at different times. With proceeding time, the colour of the solution began to turn a darker brown, indicating the formation of Pt domains on the CdSe pyramids. When the reaction was run at room temperature, this change of colour could not be observed. After quenching the reaction by removing the heat source, the hybrid nanoparticles were precipitated with methanol (1:1.5), centrifuged for 3 min at 7000 rpm and re-dispersed in 4 mL of toluene. *Note:* the supernatant is not colourless due to small amounts of Pt clusters formed during the reaction. For characterisation, aliquots were further purified by precipitation and re-dispersion with methanol (1:1), trichloromethane/acetone (1:1) and a final run with toluene/methanol (1:1), all with centrifugation for 3 min at 7000 rpm in between. Finally, the nanoparticles were re-dispersed in toluene.

5.1.5 Langmuir-Blodgett monolayer preparation and annealing

5.1.5.1 Nanoparticle purification

CdSe nanopyramids

For one film, a frozen aliquot of 1 mL CdSe pyramid reaction solution was left to warm to room temperature. The sample was separated into two centrifuge vials (2 mL) and centrifuged (5 min at 10000 rpm/9167 g). The clear brown solution was separated from the sediment, if some of the latter occurred. Then, 0.125 mL of methanol were added and the sample was centrifuged (3 min at 4500 rpm) again. After re-dispersion and precipitation with toluene and methanol (0.5/0.75 mL) the pyramids were centrifuged (3 min at 4500 rpm) and stored under nitrogen until the sub-phase for the Langmuir-Blodgett process was prepared. The pyramids were re-dispersed in 1 mL of toluene, centrifuged (3 min at 11000 rpm/ 11092 g) and precipitated from the remaining supernatant with 1 mL of methanol and centrifugation (3 min at 4500 rpm; rotation at higher values was necessary on occasion). The purified pyramids were taken up in 50 to 200 μL of toluene to give a dark brown dispersion, of which several drops were spread out onto the Langmuir-Blodgett subphase with a microlitre glass syringe. They covered roughly two thirds of the space between the barriers (set to approximately $80\,\text{cm}^2$).

Au-CdSe shell nanoparticles

Au-CdSe nanoparticles with Au shell structure at an Au/Cd ratio of 1.92 were prepared with Au(III) solution as described in section 5.1.3.3 and re-dispersed with the help of 10.1 mg (48.9 μmol) dodecanethiol. Half of a synthesis (4 mL) was mixed with ethanol (1:1), sonicated for 20 s, precipitated with centrifugation (3 min at 4500 rpm). The nanoparticles were re-dispersed in toluene (2 mL) and filtrated through a 0.45 μm PTFE syringe filter. After precipitation with ethanol (1:1) and centrifugation (3 min at 4500 rpm), the nanoparticles were re-dispersed in approximately 50 μL toluene.

Pt-CdSe nanoparticles

CdSe-Pt hybrid nanoparticles were prepared by reactions described in 5.1.4, for this purpose upscaled by a factor of two. The CdSe seed nanocrystals had been synthesised with 10 μL of DCE and were grown at 240 ± 2 °C for 4 h (Se in TOP injection at 250 °C). Particles with two sizes of Pt were prepared by letting the reaction run for five (Pt-CdSe 1.7 nm) and 24 h (Pt-CdSe 3.2 nm) and precipitated once as described above. For further purification 3/4 of the nanoparticles were first precipitated with methanol (1:1) and centrifuged at 7000 rpm for 3 min. They were taken up in the original volume of *n*-hexane and carefully cetrifuged so as not to precipitate the particles but the appearing drop-sized liquid phase (3 min at 4500 rpm). The supernatant was then transferred and precipitated with ethanol (1:1). The nanoparticles were re-dispersed in trichloromethane (5/7 of the original volume) and precipitated with acetone (1:1) and centrifugation at 7000 rpm for 3 min. Finally, the nanoparticles were re-dispersed in 50-100 µL of toluene before the Langmuir-Blodgett process.

5.1.5.2 Film preparation and annealing

Film preparation

Langmuir-Blodgett films were prepared with a KSV Nima KN-2002 system. Diethyleneglycol (DEG) was applied as subphase [297]. The trough was filled with DEG once before preparing the final subphase to remove lint and other impurities, before the nanoparticles were spread out for assembly. Silicon wafers with Au electrode structures were cleaned by bathing them in ethanol, isopropanol and demineralised water. They were then dried, mounted onto the workshop made sample holder (dipper) and brought underneath the surface of the subphase in an optimum deposition angle of $105°$ between wafer and sample holder ($15°$ between wafer and surface of subphase) [297]. TEM grids for Au-CdSe deposition were attached to silicon wafers with silver conductive paint. Compression of the monolayers was carried out by means of the producer provided software *Nima WINLB* with parameters derived from CdSe monolayer fabrication in [132]. Constant rate compression was conducted at a rate of $2\,\mathrm{mm/min}$ with a target pressure of $10\,\mathrm{N/cm^2}$ (CdSe, Pt-CdSe) or $14.5\,\mathrm{N/cm^2}$ (CdSe/Au). Once the target pressure was reached, the film was left to relax until no significant change of area was observed (around 2 h). The dipper was then lifted with a speed of $1\,\mathrm{mm/min}$. The substrates were transferred into a vacuum oven and dried over night at $<10^{-2}\,\mathrm{mbar}$. Afterwards the samples were transferred to the probe station or stored in a nitrogen flooded cabinet.

Annealing

CdSe and Pt-CdSe monolayers were annealed for 30 min under vacuum in a tubular furnace at $300\,°\mathrm{C}$. With Au-CdSe temperatures of $70\,°\mathrm{C}$, $100\,°\mathrm{C}$, $150\,°\mathrm{C}$ and $200\,°\mathrm{C}$ were applied to different samples. After cooling to room temperature under vacuum, the samples were stored in a nitrogen flooded cabinet.

5.2 Characterisation

5.2.1 Transmission Electron Microscopy (TEM), Energy Dispersive X-ray Spectroscopy (EDX)

Transmission electron microscopy was carried out with a JEOL JEM 1011 microscope with a thermal emitter operated at an accelerating voltage of 100 kV. High resolution TEM micrographs and EDX data were obtained with two TEM systems. One was a JEOL JEM 2200FS (UHR) equipped with a field emitter, CESCOR and CETCOR correctors and a Si(Li) JEOL JED-2300 energy dispersive X-ray detector at an accelerating voltage of 200 kV. Furthermore, this microscope was employed for *in-situ* heating experiments with Pt-CdSe nanoparticles employing a heatable stage. The other microscope was a Philips CM 300 UT with an EDAX DX-4 system, operated at 200 kV. Purified samples (10 µL) were drop-casted onto copper grids covered with amorphous carbon films. Medium values of nanoparticle dimensions were determined from at least 210 counts with the software *Image J*.

5.2.2 UV-Visible absorption and fluorescence spectrometry

Measurements

Absorbance measurements were carried out with a Perkin Elmer Lambda 25 two-beam and Varian Cary 50 and 500 one-beam spectrometers. Emission spectra from 500 to 750 nm were obtained with a Horiba Jobin Yvon Fluoromax-4 spectrophotometer, the irradiation wavelength was 480 nm, the slit width was set to 10 nm for CdSe nanopyramids. Emission spectra covering the wavelengths up to 1700 nm were obtained with Fluorolog-3 spectrophotometer. All measurements were conducted in quartz cuvettes with an optical path length of 10 mm. For measurements in the NIR region the samples were transferred into 1,1,2,2-tetrachloroethane by three cycles of precipitation and re-dispersion with the new solvent and acetone.

Determination of the first absorption maximum and the optical density

The wavelength of the first absorption maximum was determined at the crossing of tangents graphically fitted to the sides of the slope. The corresponding absorbance was read out as the optical density.

5.2 Characterisation

Estimation of the CdSe nanoparticle concentration

The concentration of spherical and rod-shaped CdSe nanoparticles can be calculated from the absorbance A at 350 nm. At this wavelength, the extinction coefficient ε only depends on their volume V (in cm^3)

$$\varepsilon_{350nm} = f 10^{26} V \tag{5.1}$$

with $f = 0.34$ for spherical and $f = 0.38$ for rod-shaped nanoparticles [260, 349]. The concentration c is then derived from the Lambert-Beer law

$$A = \varepsilon c d \tag{5.2}$$

with the optical path length d.

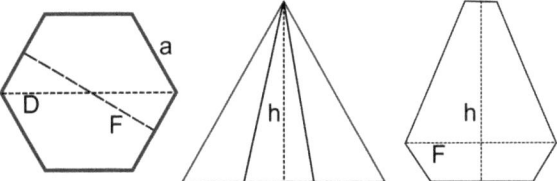

Figure 5.1: Schemes of the pyramid base, a hexagonal pyramid and a hexagonal dipyramid (left to right) with diameter D, short diameter F, edge length a and height h.

Owing to the complex geometry of nanopyramids, a calculation based on these relations contains errors but serves as an estimation. The volume was calculated from medium values of c-axis (h) and diameter (D) determined from TEM micrographs. Nanopyramids prepared with DCE were of rather dipyramidal geometry with longer c-axis. For such nanoparticles the volume may be approximated by the one of the inner ellipsoid using the short diameter F or by the volume of the inner sphere with radius $r = F/2$. For a pyramid with given dimensions the latter leads to a smaller error, so that

$$V = \frac{4}{3}\pi \left(\frac{F}{2}\right)^3 \tag{5.3}$$

was used to determine the volume. For hexagonal nanopyramids with longer diameter than c-axis, such as those prepared with COD, the formula for hexagonal pyramids

$$V = \frac{a^2}{2} h \sqrt{3} \tag{5.4}$$

was applied. In both cases $f = 0.34$ was inserted into Equation 5.1.

CHAPTER 5. EXPERIMENTAL

Determination of the quantum yield

Quantum yields (QY) were obtained by comparison of the emission signal to the one of Rhodamine 6G (95%). The absorbance of dilute samples of nanoparticles in toluene and Rhodamine 6G in ethanol was measured to determine the wavelength where the spectra crossed (set to 470-480 nm by dilution) and comparison of the integrated emission peaks obtained through excitation with the respective wavelength.

5.2.3 X-ray powder diffraction (XRD)

XRD measurements were carried out with a Philips X'Pert PRO MPD with Bragg Brentano geometry and a Cu(Kα) X-ray source emitting at 0.154 nm. Backgrounds were subtracted with the *X'Pert Highscore Plus* software. Samples were purified as stated in the respective paragraph. After the last centrifugation the nanoparticles were taken up in a few drops of toluene and drop casted onto a polished Si wafer. The diameter of crystallites D may be obtained by a re-arranged *Debye-Sherrer* formula

$$D = \frac{K\lambda 57.3}{H_B cos(\theta)} \qquad (5.5)$$

with a shape factor K (here approximated to 1) the wavelength of the incoming X-rays λ, the half-width H_B of the signal and the diffraction angle θ.

5.2.4 Scanning electron microscopy (SEM)

SEM images were obtained under vacuum with Zeiss Supra 55 and Gemini LEO 1550 systems with field emitter electron sources at an extra high tension of 5 kV.

5.2.5 Total Reflection X-ray Fluorescence Spectroscopy (TXRF)

X-ray fluorescence spectroscopy was carried out with a Bruker Picofox S2 machine with sample times of 1000 s. The samples were purified by several precipitation and re-dispersion cycles. From a solution of 0.2 mL of an aliquot from the CdSe reaction and 0.5 mL of toluene, 0.3 mL were precipitated by addition of methanol (1:1) and centrifugation (3 min at 4500 rpm), re-dispersed in hexane and centrifuged again (2 min at 4500 rpm). The supernatant was mixed with methanol/ethanol (1:1:1), centrifuged (3 min at 4500 rpm) and re-dispersed in toluene before methanol was added (1:2) and the sample was centrifuged once more (3 min at 15000 rpm). The sample taken after 10 min was purified further (hexane/ethanol 1:1, 3 min at 4500 rpm, toluene/methanol 1:1, 3 min at 4500 rpm) due to the high amount of remaining Cd-ODPA precursor (visible as slimy precipitate). For measuring, the samples were dispersed in toluene and drop-casted onto a silicon dioxide substrate.

5.2.6 Attenuated Total Reflectance Fourier Transformation Infrared Spectroscopy (ATR-FTIR)

Infrared spectra of dried samples were recorded with a Bruker Equinox 55 FTIR spectrometer with the ATR-IR technique.

5.2.7 X-ray Photoelectron Spectroscopy (XPS)

XPS measurements were carried out with the SurICat UHV XPS setup at the BESSY II storage ring, Helmholtz Zentrum Berlin. The instrument was equipped with a PM4 Plane Grating Monochromator in the beamline and a Scienta SES100 electron energy analyser in the instrument. Monochromatic X-rays with photon energies of 550 to 720 eV were applied, the energy pass was set to 50 eV for survey and 20 eV for high resolution measurements. The Au 4f signal from an Au standard sample measured in between the samples was taken as reference.

Samples not deposited *in situ* onto HOPG were purified by three cycles of precipitation and re-dispersion, as explained in the respective paragraph above (5.1.3.4), before they were drop-casted onto the freshly peeled HOPG substrates. All samples were transferred to Schlenk vials, set under vacuum to remove residues of solvents ($5*10^{-2}$ mbar) and kept under nitrogen atmosphere for storage and transport. For measurement, the substrates were attached to a metallic sample holder with conductive carbon tape.

The kinetic energy measured by the detector contains the binding energy *(BE)* of the respective electrons, the energy of the incident beam $h\nu$ and the work function of the spectrometer Φ_{spec} [350]. The binding energy can be easily obtained by

$$BE = KE - h\nu - \Phi_{spec}, \tag{5.6}$$

in which the work function of the spectrometer is obtained through reference measurements of Au.

5.2.8 Electrical transport

Interdigitated array electrode structures were prepared on n-doped silicon (100) wafers with a 300 nm oxide layer. Polymethylmetacrylate (PMMA), dissolved in chlorobenzene was spincoated onto the substrates (90 s waiting time, then rotation of 1 minute at 4000 U) and patterned by electron beam lithography with the *CAD software Elphy Quantum* from Raith by using a Zeiss Supra 55 scanning electron microscope. The dimensions of the interdigitated structure were set to 0.5 μm. With a Pfeiffer Vacuum Classic 250 vapour deposition system 2 nm Ti and 25 nm Au were deposited before the PMMA coated parts were lifted off in an acetone bath to reveal the final electrode structures. The distance between the fingers in the final electrode structures was approximately 0.45 μm. Transport measurements were carried out on a Keithley 4200-SCS semiconductor characterization system and an Agilent 4156C Precision Semiconductor Parameter Analyzer connected to a VFTTP4 probe station from Lake Shore Cryotronics. At a pressure of 10^{-5} to 10^{-6} mbar electrical fields of up to $E = 4.4 \times 10^8$ V/m (bias: 20 V) were applied. The noise level of the system is below 10 fA. Room temperature measurements were conducted with voltage steps between 0.025 and 0.5 V. For photocurrent measurements, the samples were illuminated with a CCD camera light (white light, 150 W at 21 V) or a laser with a wavelength of 637 nm and variable intensity. Irradiation wavelength dependent photocurrents were obtained by modulating the light of a Xenon lamp with a monochromator while the current was measured continuously at an applied source-drain voltage of 10 V. The measured intensity was 340 μW/cm^2 for the 560 nm component at the sample spot. Current-voltage measurements at different constant temperatures between 77.5 and 300 K were conducted with a step size of 0.2 V under liquid nitrogen cooling of the sample.

6 Summary

This thesis deals with different aspects of the fabrication and characterisation of defined metal-semiconductor hybrid nanoparticles. In the first part, systematic variations of halogenated additives in a hot-injection synthesis of CdSe nanorods enabled influencing the formation and control of hexagonal-(bi)pyramidal shapes. With the help of a semi-empirical model for crystal shapes developed by Wolff, the formation of this peculiar geometry could be traced back to an equilibrium shape forming in a thermodynamically controlled ripening stage of the reaction. Analytical and theoretical (density functional theory calculations) results supported this explanation by showing a preferential formation of Cd-rich sloped $(10\bar{1}\bar{1})$ facets by strong ligand adsorption and incorporation of halides into the ligand sphere. The type and concentration of the applied additive played a decisive role in the shape evolution. Apart from salts, the necessary halide ions can be provided by a release from organohalogen compounds. By varying the additive, a correlation between its molecular structure and the shape evolution of CdSe during the synthesis was established. The size and degree of faceting of the pyramidal nanoparticles could be controlled.

Owing to their high number of reactive sites, synthesised pyramidal nanoparticles were applied as seeds in studies on the controlled seeded-growth deposition of metals in organic solution. With Au the morphology of the formed hybrid nanoparticles was modified by changing the oxidation state of the Au precursor. Au(III)-compounds produced an amorphous Au shell which was unstable during TEM inspection and evolved into crystalline spherical domains. If Au(I)-compounds were employed, spherical domains formed instantly. Beside intensive microscopical investigations, both types of structures were examined by X-ray photoelectron spectroscopy to determine the oxidation state of the formed Au domains. Both Au(I) and Au(III) were reduced on the surface of CdSe by Se ions. Even though the major fraction of Au was in the elemental state, a smaller part of the Au dot and shell domains consisted of Au(I). The exact chemical environment of the oxidised Au species could not be determined but the position of the peak with dot-shaped domains complied with ligand bound surface atoms of Au. In accordance with literature reports on

CHAPTER 6. SUMMARY

Au nanoparticles, such unreduced surface species and the desorption of ligands are hold responsible for the migration of Au during TEM inspection.

During the attempted deposition of Ag and Pd ion exchange processes between the metals and Cd led to the formation of less defined structures. With Pd a sequential ion exchange, documented by transmission electron microscopy, energy dispersive X-ray spectroscopy and optical spectroscopy, resulted in the formation of a core-shell structure with crystalline CdSe core and an amorphous Pd_xSe_y shell.

With Pt, defined oligomeric hybrid nanoparticles with crystalline metal domains of variable sizes could be synthesised. The hybrid structures were stable against electron beams and did not change significantly during *in-situ* annealing in an electron microscope. Due to the size tunability of the metal domains and their stability, these nanoparticles are suitable for further investigations and possible future applications in the realm of (opto)electronics.

In the last step, Pt-CdSe hybrid nanoparticles were assembled into two-dimensional monolayers and tested for their electrical and photoelectrical properties in comparison to bare CdSe. With metal domains of 3.2 nm an increase in conductance of up to nine orders of magnitude was recorded in darkness. Room temperature current-voltage curves were affected by barrier lowering processes. Contrary to the clear increase of conductivity under white light irradiation of CdSe arrays, laser radiation was necessary to observe a difference between dark and photocurrents with hybrid nanoparticles. With small Pt domains of 1.7 nm the measured current-voltage curves were similar to those of pure CdSe. This implies the possibility of increasing and varying the conductivity of semiconductor array by controlled formation of hybrid nanoparticles with variable sizes of metal domains.

The method of influencing the semiconductor shape evolution by addition of varied halogenated compounds is expected to be transferable to other II-VI compounds with wurtzite structure. This way, the application of additives releasing surface active species during the synthesis may enrich the tool-box of colloidal chemistry and allow for the fabrication of new morphologies. The studies on the formation and instability of an Au shell on CdSe nanoparticles contribute to the understanding of metal deposition in general and may help to explain occurring instabilities in other hybrid nanoparticles. The investigations on electrical transport are basic studies that indicate the applicability of the system and provide guidance for planned projects dealing with the underlying transport mechanisms.

7 Zusammenfassung

Im Rahmen der vorliegenden Arbeit wurden verschiedene Aspekte der Herstellung und Charakterisierung von hybriden Metall-Halbleiter Nanostrukturen untersucht. Im ersten Teil wurde durch eine systematische Variation von halogenhaltigen Additiven Einfluß auf die Entstehung und Kontrolle von CdSe-Nanopartikeln in hexagonal-(bi-)pyramidaler Form gewonnen. Die Ausbildung dieser für die Hybridbildung sehr interessanten Form konnte mit Hilfe eines von Wolff entwickelten Modells auf Gleichgewichtsformen mit niedriger Oberflächenenergie zurückgeführt werden, die durch thermodynamisch kontrolliertes Reifen der Nanopartikel entstehen. Hierbei werden schräge, cadmiumreiche $(10\bar{1}\bar{1})$ Facetten in CdSe durch starke Ligandenstabilisierung und den Einbau von Halogeniden in die Ligandensphäre begünstigt, was anhand experimenteller und theoretischer Befunde (Simulationen mittels Dichtefunktionaltheorie) belegt wurde. Die Art und Konzentration der Additive spielte hierbei eine besondere Rolle. Neben halogenidhaltigen Salzen können auch hinzugefügte Organohalogenverbindungen während der Synthese Halogenide freisetzen. Durch die Verwendung unterschiedlicher Halogenalkane konnte ein Zusammenhang zwischen der Formentwicklung der Nanopartikel im Laufe einer *Hot-Injection*-Synthese und der Art und Struktur der Halogenverbindung hergestellt werden. Hierbei wurde die Größe der Nanopartikel variiert und die Ausprägung der Facettierung beeinflusst.

Hergestellte hexagonal-pyramidalen Partikel wurden aufgrund ihrer besonderen Oberflächenbeschaffenheit mit vielen exponierten Positionen als Keime in Studien zur Herstellung von definierten Hybridnanopartikeln mit unterschiedlichen Metallen verwendet. Es wurde mit einer Keimwachstumsmethode in organischer Lösung gearbeitet. Bei der Kombination von CdSe und Au konnte die Morphologie der gebildeten Hybridnanopartikel in Abhängigkeit von der Oxidationsstufe der eingesetzten Goldverbindung beeinflusst werden. Mit Au(III)-Verbindungen wurde ein ungewöhnliches Abscheidungsverhalten in Form der Bildung einer größtenteils amorphen und gegenüber Elektronenstrahlen instabilen Schale auf der Oberfläche von CdSe beobachtet. Definierte sphärische Au-Domänen konnten mit Au(I)- statt Au(III)-Verbindungen erhalten werden. Nanopartikel beider Typen wurden mittels Synchrotron-Röntgenphotoelektronenspektroskopie und elektronenmikroskopischer

CHAPTER 7. ZUSAMMENFASSUNG

Methoden untersucht. Dabei wurde gezeigt, dass Au(III) an der Oberfläche der Partikel von Se-Ionen reduziert wird. Ein Anteil an nichtreduziertem Au befindet sich jedoch in der Schalenstruktur, was in Kombination mit der Wechselwirkung von Elektronenstrahl und Liganden eine Erklärung des beschriebenen Verhaltens liefert. Auch die sphärischen Abscheidungen enhielten einen Anteil von nicht vollständig reduziertem Au, was mit der Bindung von Oberflächenatomen an Liganden erklärt wurde. Bei der Abscheidung von Ag und Pd kam es aufgrund von Ionenaustauschprozessen zwischen den Metallionen zur Ausbildung weniger definierter Strukturen. Mit Pd entstand durch sequentiellen Ionenaustausch innerhalb der Partikel, nachgewiesen durch Transmissionselektronenmikroskopie, Energiedispersive Röntgenspektroskopie und optische Spektroskopie, eine Kern-Schale-Struktur aus einem kristallinen CdSe Kern und einer amorphen-Pd_xSe_y Schale.

Mit Pt konnten definierte Hybridnanopartikel mit größenvariierbaren sphärischen Pt-Domänen hergestellt werden. In elektronenmikroskopischen und röntgenographischen Untersuchungen konnte die hohe Kristallinität der Metallabscheidungen nachgewiesen werden. Sie zeigten keine Instabilitäten gegenüber dem Elektronenstrahl und veränderten sich auch beim *in-situ* Tempern im Elektronenmikroskop wenig. Durch die strukturelle Stabilität eignen sich die Partikel für weitergehende Untersuchungen und perspektivisch auch Anwendungen im Bereich der (Opto-)elektronik.

Pt-CdSe-Hybridnanopartikel wurden schließlich als Monolage im Vergleich zu reinen CdSe-Nanopartikeln auf ihre Transporteigenschaften für elektrischen Strom und deren Photoabhängigkeit untersucht. Mit 3.2 nm großem Pt wurde eine Steigerung der Leitfähigkeit um bis zu neun Zehnerpotenzen im Dunkeln beobachtet. Die Strom-Spannungskennlinien zeigten eine Widerstandsreduktion durch den Metall-Halbleiterkontakt. Während reine CdSe-Nanopartikel einen deutlichen Anstieg der Leitfähigkeit durch Beleuchtung zeigten, war dies für Hybridnanopartikel mit Pt aufgrund des bereits hohen Dunkelstroms nur unter Laserbestrahlung nachweisbar. Mit kleineren Pt-Domänen (1.7 nm) ähnelte die Stromstärke derjenigen, die durch reine CdSe Proben floss. Dies deutet an, dass Pt die Charakteristik stark beeinflusst und es möglich ist die Leitfähigkeit von Halbleiternanopartikeln durch definiert aufgewachsene Metalldomänen zu erhöhen und durch die Größe der Domänen zu variieren.

Die im Rahmen der Arbeit gewonnenen Erkenntnisse zur Formkontrolle von Halbleiternanopartikeln sollten auf andere II-VI-Halbleiter mit Wurtzitstruktur übertragbar sein. Dadurch hat die Zugabe von halogenhaltigen Additiven, die während der Synthese oberflächenaktive Halogenide freisetzen, das Potenzial die Kolloidchemie um eine Methode zur Erschließung neuer Nanopartikelgeometrien zu bereichern. Die Untersuchungen zu Bildung und Instabilität der Au-Schale auf pyramidalen CdSe-Nanopartikeln leisten einen Beitrag zum Verständnis der Abscheidungsreaktion und allgemein auftretender Instabilitäten in diesem System. Die Ergebnisse der Transportstudien an Pt-CdSe-Strukturen zeigen als Grundlagenexperimente Richtungen für weitere geplante Studien zur Untersuchung der vorliegenden Mechanismen und Potenziale des Systems auf. Hierunter sind Experimente zu etwaigen Transistorcharakteristiken, die durch die Größe der Pt-Domänen über den Coulombeffekt gesteuert werden könnten.

A Additional data

Additional data from chapter 2

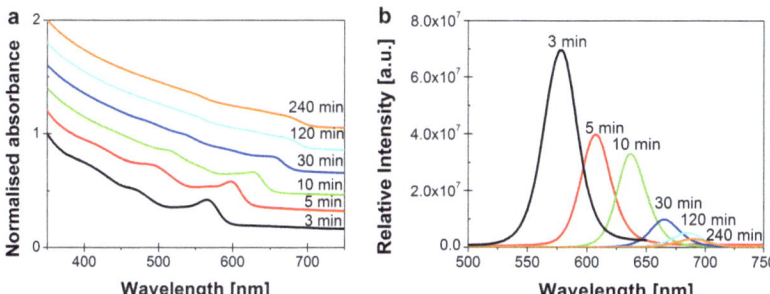

Figure A.1: (a) Temporal evolution of absorbance and (b) emission spectra of aliquots from a reaction with DCE as additive. The absorbance spectra were normalised to 1 at 350 nm. The same factor was then applied to the emission to compare the relative intensities.

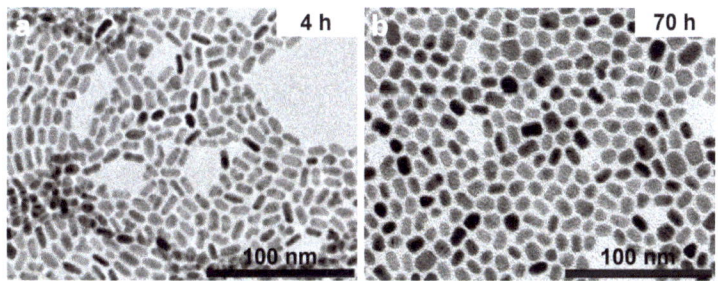

Figure A.2: TEM micrographs of samples from control reactions after four and 70 hours. As expected from Ostwald ripening the long reaction time lead to inhomogeneous shapes and a large size distribution.

APPENDIX A. ADDITIONAL DATA

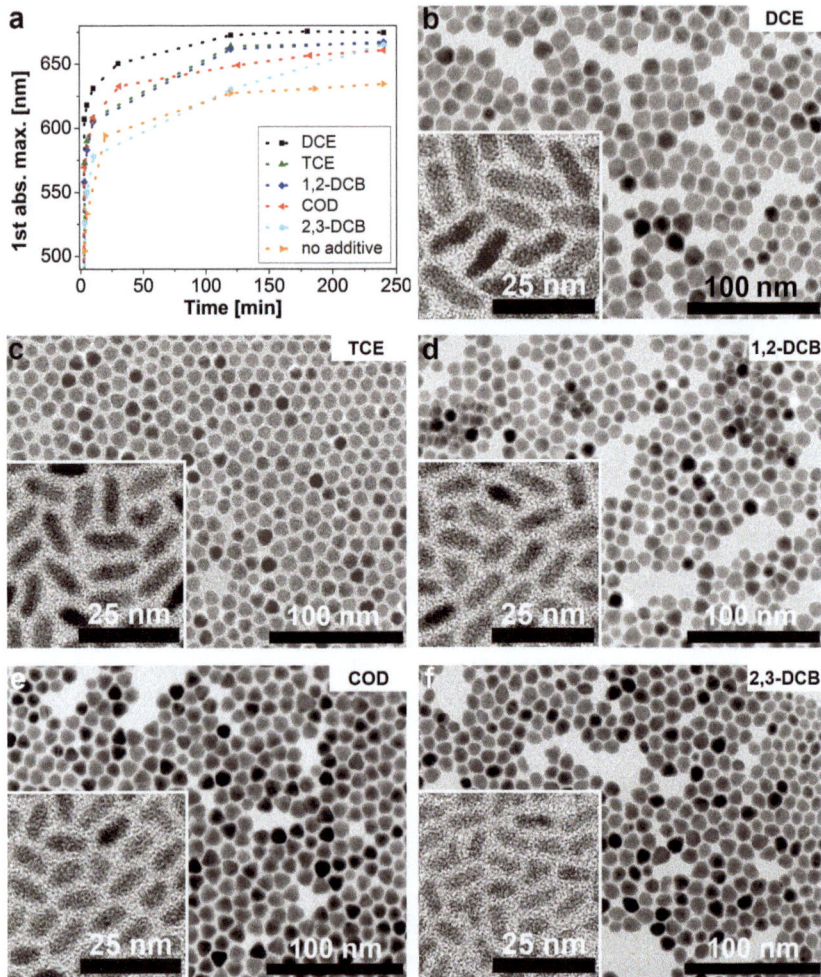

Figure A.3: (a) Temporal evolution of the first absorption maximum and (b - f) TEM micrographs of samples prepared with 1,2-dichloroethane (DCE), 1,1,2-trichloroethane (TCE), 1,2-dichlorobutane (1,2-DCB), 1-chlorooctadecane (COD) and 2,3-dichlorobutane after 4 hours and after 10 minutes (insets).

Additional data from chapter 3

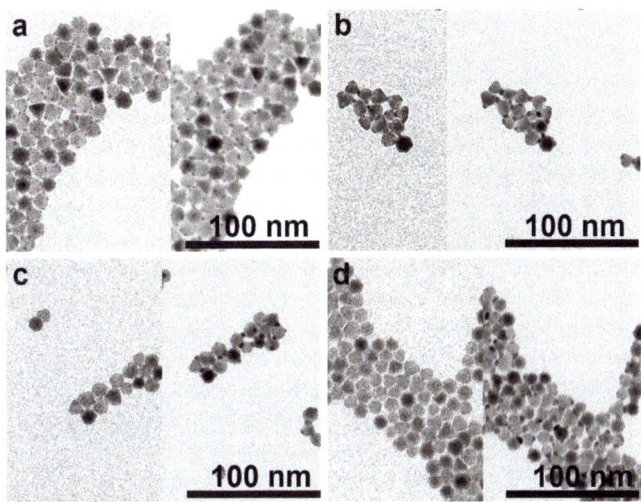

Figure A.4: Electron beam induced evolution of samples examined by EDX and XPS in Chapter 3, Section 3.2.1. The time between the samples fom left to right was approximately 150 s, the beam current ranged between 7 and 11 pA/cm^2. (a) Au-CdSe dots with Au/CdSe 1.40, no significant changes are observed. (b) In Au-CdSe dots samples with Au/CdSe 1.97 slight growth can be seen. (c) and (d) are Au-shell samples with Au/CdSe 1.97 and 2.88, were Au migration is clearly visible.

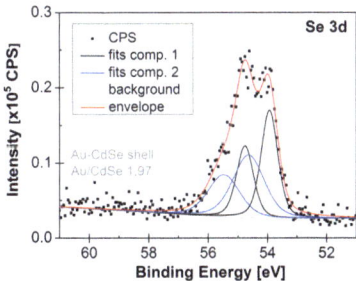

Figure A.5: XPS Se 3d signal of a Au-CdSe shell sample with Au/CdSe 1.97. Same as the peaks shown in the main text, it can be fitted with two components with one Se 3d 5/2 bulk contribution at 53.9 eV and a broader contribution at 54.6 eV

APPENDIX A. ADDITIONAL DATA

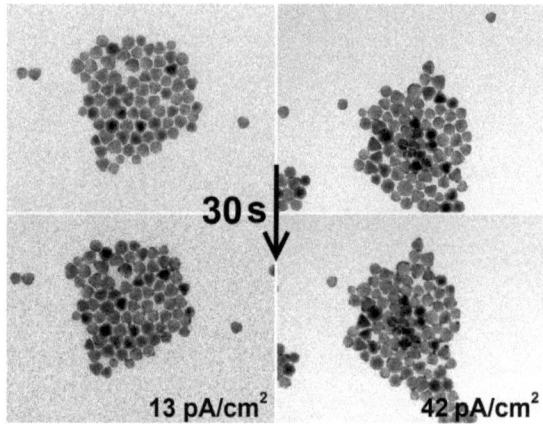

Figure A.6: The influence of the beam intensity on the changes in an Au-CdSe sample is shown by a comparison of the temporal evolution under varied beam currents at two different spots on the same grid but well separated in position. During 30 seconds, the growth of Au domains has proceeded more clearly at a beam current of $42\,\mathrm{pA/cm^2}$ than with $13\,\mathrm{pA/cm^2}$.

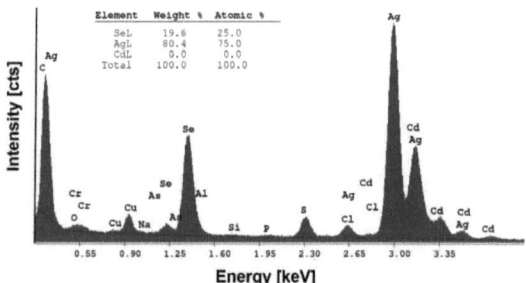

Figure A.7: EDX measurement of nanoparticles after ion exchange from Cd(II) to Ag(I).

IV

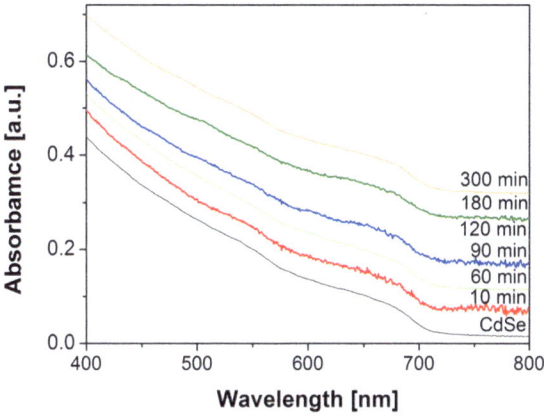

Figure A.8: Absorbance spectra of CdSe nanopyramids incubated with palladium acetate at room temperature after different periods of time.

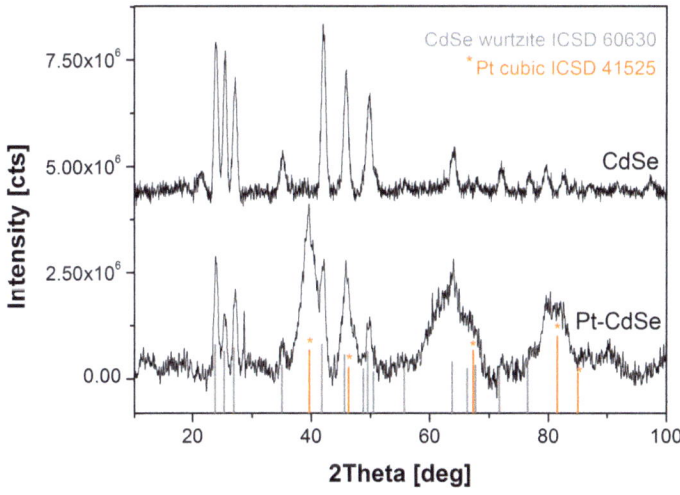

Figure A.9: X-ray diffraction patterns of CdSe and Pt-CdSe nanoparticles (24 h). Reference data: CdSe JCPDS # 00-008-0459, [60]; Pt JCPDS # 01-088-2343, [236].

APPENDIX A. ADDITIONAL DATA

Additional data from chapter 4

Figure A.10: Electrode with CdSe nanopyramid array after annealing.

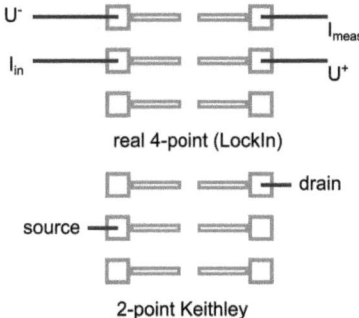

Figure A.11: In a 4-point probe measurement, the current is recorded between electrodes different to those between which the voltage is applied. In the standard (Keithley) set-up only two electrodes are employed. This figure appears under courtesy of H. Lehmann.

B Safety

Safety information for employed substances

Table B.1: Substances and corresponding acronyms for hazard symbols, signal words, H-statements and P-statement. The information stems from reference [351] if not indicated else.

Substance	Hazard pictogram & signal word	H-and EUH-statements	P-statements
Acetone	02, 07 Danger	H225-H319-H336, EUH066	P210-P233-P305+P351+P338
Cadmium chloride	06, 08, 09 Danger	H350-H340-H360FD-H330-H301-H372-H410	P260-P284-P301-P320-P405-P501
Cadmium oxide	06, 08, 09 Danger	H350-H341-H361fd-H330-H372-H410	P201-P281-P273-P308+P313-P304+P340-P309+P310
1-Chlorooctadecane [352]	07 Warning	H315-H319-H335	P261-P305+P351+P338
1,2-Dibromoethane [353]	06, 08, 09 Danger	H315-H319-H350-H411-H311-H331-H335-H301	P280-P302+P350-P308+P313-P304+P340-P261-P273-P305+P351+P338

APPENDIX B. SAFETY

Substance	Hazard pictogram & signal word	H- and EUH- statements	P-statements
1,2-Dichlorobutane [354]	02, 07 Warning	H226-H315-H319-H335	P261-P305+P351+P338
2,3-Dichlorobutane	02 Danger	H225	P210-P240
1,2-Dichloroethane [355]	02, 07, 08 Danger	H225-H350-H302-H319-H335-H315	P201-P210-P302+P352-P304+P340-P305+P351+P338-P308+P313
1-Chlorooctadecane [352]	07 Warning	H315-H319-H335	P261-P305+P351+P338
1,2-Dibromoethane [353]	06, 08, 09 Danger	H315-H319-H350-H411-H311-H331-H335-H301	P280-P302+P350-P308+P313-P304 + P340-P261-P273-P305+P351+P338
Diethylene glycol	07 Danger	H302	
1,2-Diiodoethane [356]	07 Warning	H315-H319-H335	P261-P305+P351+P338
1-Dodecanethiol	02, 07 Warning	H335-H315-H319	P302+P352-P304+P340-P305+P351+P338
1-Dodecylamine	05, 07, 09 Danger	H302-H314-H400-H290	P280-P273-P301+P330+P331-P305+P351+P338-P309+P310
n-Dodecyltrimethyl-ammonium bromide [357]	06, 09 Danger	H301-H315-H319-H335-H410	P301+P310-P305+P351+P338-P302+P352-P321-P405-P501

Substance	Hazard pictogram & signal word	H- and EUH-statements	P-statements
n-Dodecyltrimethyl-ammonium chloride [358]	07, 09 Warning	H302-H315-H319-H410-H335	P261-P273-P301+P312-P302+P352-P280-P305+P351+P338
Ethanol	02 Danger	H225	P210
Gold(III) chloride	05 Danger	H314	P260-P301+P330+P331-P303+P361+P353-P305+P351+P338-P405-P501
Hexadecylamine	05, 09 Danger	H302-H314-H410	P280-P273-P301+P330+P331-P305+P351+P338-P309+P310
n-hexane	02, 07, 08, 09 Danger	H225-H304-H361f-H373-H315-H336-H441	P210-P240-P273-P301+P310-P331 P302+P352-P403+P235
Methanol	02, 06, 08 Danger	H225-H331-H311f-H301-H370	P210-P233-P280-P302+P352-P309+P310
cis-9-Octadecenyl-amine/oleylamine	05, 07, 08, 09 Danger	H225-H304-H361f-H373-H315-H336-H441	P210-P240-P273-P301+P310-P331 P302+P352-P403+P235
n-Octadecylphosphonic acid [359]	07 Warning	H315-H319-H335	P261-P280-P305+P351+P338-P304+P340-P405-P501

APPENDIX B. SAFETY

Substance	Hazard pictogram & signal word	H- and EUH- statements	P-statements
Oleic acid	Not a dangerous substance according to GHS. Regulation (EC) No 1272/2008.		
Palladium(II) acetate [360]	07 Warning	H319	P280- P305+P351+P338
Palladium(II) acetylacetonate	05 Danger	H314	P260- P301+P330+P331
Platinum(II) acetylacetonate [361]	07, 08 Warning	H302-H312-H332- H315-H319-H335- H361	P261-P280- P305+P351+P338
2-Propanol	02, 07 Danger	H225-H319-H336	P210-P233-P273- P305+P351+P338
Selenium	06, 08 Danger	H301-H331-H373- H413	P260-P273- P301+P310- P304+P340
Silver(I) acetate	07, 09 Warning	H315-H319-H335- H400	P261-P273- P305+P351+P338
Toluene	02, 07, 08 Danger	H225-H304- H361d- H373-H315-H336	P210-P301+P310- P311-P302+P352
1,1,2-Trichloroethane	07, 08 Warning	H302-H312-H32- H351-EUH066	P280
Trichloromethane	06, 08 Danger	H351-H361d- H331- H302-H372-H319- H315	P301+P310- P305+P351+P338- P302+P352- P321-P405-P501
Tri-n-octylphosphane [362]	05 Danger	H314	P280- P305+P351+P338- P310
Tri-n-octylphosphane oxide [363]	07, 09 Warning	H319-H411	P262-P273- P305+P351+P338

Health (H), Supplemental Hazard Information (EUH-statements) and precautionary (P) statements (GHS)

Table B.2: All H, EUH, and P Statements.

Identifier	Statement
H200	Unstable explosives.
H201	Explosive; mass explosion hazard.
H202	Explosive, severe projection hazard.
H203	Explosive; fire, blast or projection hazard.
H204	Fire or projection hazard.
H205	May mass explode in fire.
H220	Extremely flammable gas.
H221	Flammable gas.
H222	Extremely flammable aerosol.
H223	Flammable aerosol.
H224	Extremely flammable liquid and vapour.
H225	Highly flammable liquid and vapour.
H226	Flammable liquid and vapour.
H228	Flammable solid.
H240	Heating may cause an explosion.
H241	Heating may cause a fire or explosion.
H242	Heating may cause a fire.
H250	Catches fire spontaneously if exposed to air.
H251	Self-heating: may catch fire.
H252	Self-heating in large quantities; may catch fire.
H260	In contact with water releases flammable gases which may ignite spontaneously.

continues on next page

APPENDIX B. SAFETY

Identifier	Statement
H261	In contact with water releases flammable gases.
H270	May cause or intensify fire; oxidiser.
H271	May cause fire or explosion; strong oxidiser.
H272	May intensify fire; oxidiser.
H280	Contains gas under pressure; may explode if heated.
H281	Contains refrigerated gas; may cause cryogenic burns or injury.
H290	May be corrosive to metals.
H300	Fatal if swallowed.
H301	Toxic if swallowed.
H302	Harmful if swallowed.
H304	May be fatal if swallowed and enters airways.
H310	Fatal in contact with skin.
H311	Toxic in contact with skin.
H312	Harmful in contact with skin.
H314	Causes severe skin burns and eye damage.
H315	Causes skin irritation.
H317	May cause an allergic skin reaction.
H318	Causes serious eye damage.
H319	Causes serious eye irritation.
H330	Fatal if inhaled.
H331	Toxic if inhaled.
H332	Harmful if inhaled.
H334	May cause allergy or asthma symptoms orbreathing difficulties if inhaled.
H335	May cause respiratory irritation.
H336	May cause drowsiness or dizziness.
H340	May cause genetic defects.

continues on next page

Identifier	Statement
H341	Suspected of causing genetic defects.
H350	May cause cancer.
H351	Suspected of causing cancer.
H360	May damage fertility or the unborn child.
H361	Suspected of damaging fertility or the unborn child.
H362	May cause harm to breast-fed children.
H370	Causes damage to organs.
H371	May cause damage to organs.
H372	Causes damage to organs through prolonged or repeated exposure.
H373	May cause damage to organs through prolonged or repeated exposure.
H400	Very toxic to aquatic life.
H410	Very toxic to aquatic life with long lasting effects.
H411	Toxic to aquatic life with long lasting effects.
H412	Harmful to aquatic life with long lasting effects.
H413	May cause long lasting harmful effects to aquatic life.
H350i	May cause cancer by inhalation.
H360F	May damage fertility.
H360D	May damage the unborn child.
H361f	Suspected of damaging fertility.
H361d	Suspected of damaging the unborn child.
H360FD	May damage fertility. May damage the unborn child.
H361fd	Suspected of damaging fertility. Suspected of damaging the unborn child.
H360Fd	May damage fertility. Suspected of damaging the unborn child.

continues on next page

APPENDIX B. SAFETY

Identifier	Statement
H360Df	May damage the unborn child. Suspected of damaging fertility.
EUH001	Explosive when dry.
EUH006	Explosive with or without contact with air.
EUH014	Reacts violently with water.
EUH018	In use may form flammable/explosive vapour-air mixture.
EUH019	May form explosive peroxides.
EUH044	Risk of explosion if heated under confinement.
EUH029	Contact with water liberates toxic gas.
EUH031	Contact with acids liberates toxic gas.
EUH032	Contact with acids liberates very toxic gas.
EUH066	Repeated exposure may cause skin dryness or cracking.
EUH070	Toxic by eye contact.
EUH071	Corrosive to the respiratory tract.
EUH059	Hazardous to the ozone layer.
EUH201	Contains lead. Should not be used on surfaces liable to be chewed or sucked by children.
EUH201A	Warning! contains lead.
EUH202	Cyanoacrylate. Danger. Bonds skin and eyes in seconds. Keep out of the reach of children.
EUH203	Contains chromium (VI). May produce an allergic reaction.
EUH204	Contains isocyanates. May produce an allergic reaction.
EUH205	Contains epoxy constituents. May produce an allergic reaction.
EUH206	Warning! Do not use together with other products. May release dangerous gases (chlorine).
EUH207	Warning! Contains cadmium. Dangerous fumes are formed during use. See information supplied by the manufacturer. Comply with the safety instructions.

continues on next page

Identifier	Statement
EUH208	Contains <name of sensitising substance>. May produce an allergic reaction.
EUH209	Can become highly flammable in use.
EUH209A	Can become flammable in use.
EUH210	Safety data sheet available on request.
EUH401	To avoid risks to human health and the environment, comply with the instructions for use.
P101	If medical advice is needed, have product container or label at hand.
P102	Keep out of reach of children.
P103	Read label before use.
P201	Obtain special instructions before use.
P202	Do not handle until all safety precautions have been read and understood.
P210	Keep away from heat/sparks/open flames/hot surfaces. — No smoking.
P211	Do not spray on an open flame or other ignition source.
P220	Keep/Store away from clothing/.../combustible materials.
P221	Take any precaution to avoid mixing with combustibles ...
P222	Do not allow contact with air.
P223	Keep away from any possible contact with water, because of violent reaction and possible flash fire.
P230	Keep wetted with ...
P231	Handle under inert gas.
P232	Protect from moisture.
P233	Keep container tightly closed.
P234	Keep only in original container.
P235	Keep cool.

continues on next page

APPENDIX B. SAFETY

Identifier	Statement
P240	Ground/bond container and receiving equipment.
P241	Use explosion-proof electrical/ventilating/lighting/... equipment.
P242	Use only non-sparking tools.
P243	Take precautionary measures against static discharge.
P244	Keep reduction valves free from grease and oil.
P250	Do not subject to grinding/shock/.../friction.
P251	Pressurized container: Do not pierce or burn, even after use.
P260	Do not breathe dust/fume/gas/mist/vapours/spray.
P261	Avoid breathing dust/fume/gas/mist/vapours/spray.
P262	Do not get in eyes, on skin, or on clothing.
P263	Avoid contact during pregnancy/while nursing.
P264	Wash ... thoroughly after handling.
P270	Do not eat, drink or smoke when using this product.
P271	Use only outdoors or in a well-ventilated area.
P272	Contaminated work clothing should not be allowed out of the workplace.
P273	Avoid release to the environment.
P280	Wear protective gloves/protective clothing/eye protection/face protection.
P281	Use personal protective equipment as required.
P282	Wear cold insulating gloves/face shield/eye protection.
P283	Wear fire/flame resistant/retardant clothing.
P284	Wear respiratory protection.
P285	In case of inadequate ventilation wear respiratory protection.
P231 + P232	Handle under inert gas. Protect from moisture.
P235 + P410	Keep cool. Protect from sunlight.
P301	IF SWALLOWED:

continues on next page

Identifier	Statement
P302	IF ON SKIN:
P303	IF ON SKIN (or hair):
P304	IF INHALED:
P305	IF IN EYES:
P306	IF ON CLOTHING:
P307	IF exposed:
P308	IF exposed or concerned:
P309	IF exposed or if you feel unwell:
P310	Immediately call a POISON CENTER or doctor/physician.
P311	Call a POISON CENTER or doctor/physician.
P312	Call a POISON CENTER or doctor/physician if you feel unwell.
P313	Get medical advice/attention.
P314	Get medical advice/attention if you feel unwell.
P315	Get immediate medical advice/attention.
P320	Specific treatment is urgent (see ... on this label).
P321	Specific treatment (see ... on this label).
P322	Specific measures (see ... on this label).
P330	Rinse mouth.
P331	Do NOT induce vomitting.
P332	If skin irritation occurs:
P333	If skin irritation or rash occurs:
P334	Immerse in cool water/wrap in wet bandages.
P335	Brush off loose particles from skin.
P336	Thaw frosted parts with lukewarm water. Do not rub affected area.
P337	If eye irritation persists:

continues on next page

APPENDIX B. SAFETY

Identifier	Statement
P338	Remove contact lenses, if present and easy to do. Continue rinsing.
P340	Remove victim to fresh air and keep at rest in a position comfortable for breathing.
P341	If breathing is difficult, remove victim to fresh air and keep at rest in a position comfortable for breathing.
P342	If experiencing respiratory symptoms:
P350	Gently wash with plenty of soap and water.
P351	Rinse cautiously with water for several minutes.
P352	Wash with plenty of soap and water.
P353	Rinse skin with water/shower.
P360	Rinse immediately contaminated clothing and skin with plenty of water before removing clothes.
P361	Remove/Take off immediately all contaminated clothing.
P362	Take off contaminated clothing and wash before reuse.
P363	Wash contaminated clothing before reuse.
P370	In case of fire:
P371	In case of major fire and large quantities:
P372	Explosion risk in case of fire.
P373	DO NOT fight fire when fire reaches explosives.
P374	Fight fire with normal precautions from a reasonable distance.
P375	Fight fire remotely due to the risk of explosion.
P376	Stop leak if safe to do so.
P377	Leaking gas fire: Do not extinguish, unless leak can be stopped safely.
P378	Use ... for extinction.
P380	Evacuate area.
P381	Eliminate all ignition sources if safe to do so.

continues on next page

Identifier	Statement
P390	Absorb spillage to prevent material damage.
P391	Collect spillage.
P301 + P310	IF SWALLOWED: Immediately call a POISON CENTER or doctor/physician.
P301 + P312	IF SWALLOWED: Call a POISON CENTER or doctor/physician if you feel unwell.
P301 + P330 + P331	IF SWALLOWED: rinse mouth. Do NOT induce vomitting.
P302 + P334	IF ON SKIN: Immerse in cool water/wrap in wet bandages.
P302 + P350	IF ON SKIN: Gently wash with plenty of soap and water.
P302 + P352	IF ON SKIN: Wash with plenty of soap and water.
P303 + P361 + P353	IF ON SKIN (or hair): Remove/Take off immediately all contaminated clothing. Rinse skin with water/shower.
P304 + P340	IF INHALED: Remove victim to fresh air and keep at rest in a position comfortable for breathing.
P304 + P341	IF INHALED: If breathing is difficult, remove victim to fresh air and keep at rest in a position comfortable for breathing.
P305 + P351 + P338	IF IN EYES: Rinse cautiously with water for several minutes. Remove contact lenses, if present and easy to do. Continue rinsing.
P306 + P360	IF ON CLOTHING: Rinse immediately contaminated clothing and skin with plenty of water before removing clothes.
P307 + P311	IF exposed: Call a POISON CENTER or doctor/physician.
P308 + P313	IF exposed or concerned: Get medical advice/attention.
P309 + P311	IF exposed or if you feel unwell: Call a POISON CENTER or doctor/physician.
P332 + P313	If skin irritation occurs: Get medical advice/attention.
P333 + P313	If skin irritation or rash occurs: Get medical advice/attention.
P335 + P334	Brush off loose particles from skin. Immerse in cool water/wrap in wet bandages.

continues on next page

APPENDIX B. SAFETY

Identifier	Statement
P337 + P313	If eye irritation persists: Get medical advice/attention.
P342 + P311	If experiencing respiratory symptoms: Call a POISON CENTER or doctor/physician.
P370 + P376	In case of fire: Stop leak if safe to do so.
P370 + P378	In case of fire: Use ... for extinction.
P370 + P380	In case of fire: Evacuate area.
P370 + P380 + P375	In case of fire: Evacuate area. Fight fire remotely due to the risk of explosion.
P371 + P380 + P375	In case of major fire and large quantities: Evacuate area. Fight fire remotely due to the risk of explosion.
P401	Store ...
P402	Store in a dry place.
P403	Store in a well-ventilated place.
P404	Store in a closed container.
P405	Store locked up.
P406	Store in corrosive resistant/... container with a resistant inner liner.
P407	Maintain air gap between stacks/pallets.
P410	Protect from sunlight.
P411	Store at temperatures not exceeding °C/°F.
P412	Store at temperatures not exceeding 50 °C/122 °F.
P413	Store bulk masses greater than kg/lbs at temperatures not exceeding °C/°F.
P420	Store away from other materials.
P422	Store contents under ...
P402 + P404	Store in a dry place. Store in a closed container.
P403 + P233	Store in a well-ventilated place. Keep container tightly closed.
P403 + P235	Store in a well-ventilated place. Keep cool.

continues on next page

Identifier	Statement
P410 + P403	Protect from sunlight. Store in a well-ventilated place.
P410 + P412	Protect from sunlight. Do not expose to temperatures exceeding 50 °C/122 °F.
P411 + P235	Store at temperatures not exceeding °C/°F. Keep cool.
P501	Dispose of contents/container to ...

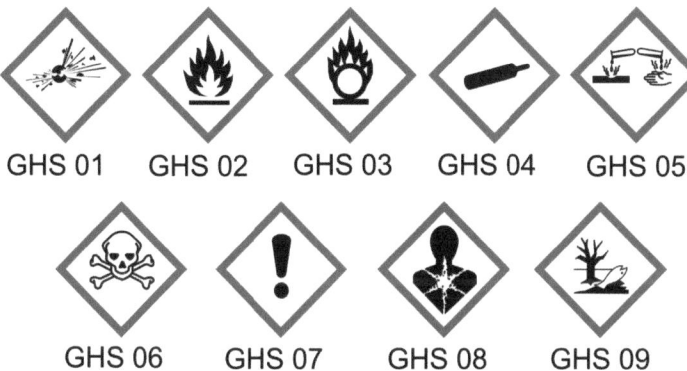

Figure B.1: GHS-pictograms. GHS01–explosive, GHS02–flammable, GHS03–oxidising, GSH04–compressed gas, GHS05–corrosive, GHS06–toxic, GHS07 irritant, GHS08–health hazard, GHS09–environmentally damaging.

Bibliography

[1] M. B. Wilker, K. J. Schnitzenbaumer, G. Dukovic. Recent Progress in Photocatalysis Mediated by Colloidal II-VI Nanocrystals. *Isr. J. Chem.* **2012**, *52*, 1002–1015.

[2] Y. Li, G. A. Somorjai. Nanoscale Advances in Catalysis and Energy Applications. *Nano Lett.* **2010**, *10*, 2289–2295.

[3] D. V. Talapin, J.-S. Lee, M. V. Kovalenko, E. V. Shevchenko. Prospects of Colloidal Nanocrystals for Electronic and Optoelectronic Applications. *Chem. Rev.* **2010**, *110*, 389–458.

[4] D. R. Cooper, J. L. Nadeau. Nanotechnology for in vitro neuroscience. *Nanoscale* **2009**, *1*, 183–200.

[5] U. I. Tromsdorf, O. T. Bruns, S. C. Salmen, U. Beisiegel, H. Weller. A Highly Effective, Nontoxic T1 MR Contrast Agent Based on Ultrasmall PEGylated Iron Oxide Nanoparticles. *Nano Lett.* **2009**, *9*, 4434–4440.

[6] H. Weller. Colloidal Semiconductor Q-Particles: Chemistry in the Transition Region Between Solid State and Molecules. *Angew. Chem. Int. Ed.* **1993**, *32*, 41–53.

[7] C. B. Murray, D. J. Norris, M. G. Bawendi. Synthesis and characterization of nearly monodisperse CdE (E = sulfur, selenium, tellurium) semiconductor nanocrystallites. *J. Am. Chem. Soc.* **1993**, *115*, 8706–8715.

[8] R. Costi, A. Saunders, U. Banin. Colloidal Hybrid Nanostructures: A New Type of Functional Materials. *Angew. Chem. Int. Ed.* **2010**, *49*, 4878–4897.

[9] C. Pacholski, A. Kornowski, H. Weller. Site-Specific Photodeposition of Silver on ZnO Nanorods. *Angew. Chem.* **2004**, *116*, 4878–4881.

[10] T. Mokari, E. Rothenberg, I. Popov, R. Costi, U. Banin. Selective Growth of Metal Tips onto Semiconductor Quantum Rods and Tetrapods. *Science* **2004**, *304*, 1787–1790.

[11] R. Costi, A. E. Saunders, E. Elmalem, A. Salant, U. Banin. Visible Light-Induced Charge Retention and Photocatalysis with Hybrid CdSe-Au Nanodumbbells. *Nano Lett.* **2008**, *8*, 637–641.

[12] K. Maeda. Photocatalytic water splitting using semiconductor particles: History and recent developments. *J. Photochem. Photobiol. C* **2011**, *12*, 237–268.

[13] M. T. Sheldon, P.-E. Trudeau, T. Mokari, L.-W. Wang, A. P. Alivisatos. Enhanced Semiconductor Nanocrystal Conductance via Solution Grown Contacts. *Nano Lett.* **2009**, *9*, 3676–3682.

[14] C. Burda, X. Chen, R. Narayanan, M. A. El-Sayed. Chemistry and Properties of Nanocrystals of Different Shapes. *Chem. Rev.* **2005**, *105*, 1025–1102.

[15] B. H. Juárez, C. Klinke, A. Kornowski, H. Weller. Quantum Dot Attachment and Morphology Control by Carbon Nanotubes. *Nano Lett.* **2007**, *7*, 3564–3568.

[16] B. H. Juárez, M. Meyns, A. Chanaewa, Y. Cai, C. Klinke, H. Weller. Carbon Supported CdSe Nanocrystals. *J. Am. Chem. Soc.* **2008**, *130*, 15282–15284.

[17] M. Meyns, *Synthesis of hybrid cadmium selenide-gold nanostructures*, Diplomarbeit, Fachbereich Chemie, Universität Hamburg, **2010**.

[18] P. V. Kamat. Quantum Dot Solar Cells. The Next Big Thing in Photovoltaics. *J. Phys. Chem. Lett.* **2013**, *4*, 908–918.

[19] I. J. Kramer, E. H. Sargent. The Architecture of Colloidal Quantum Dot Solar Cells: Materials to Devices. *Chem. Rev.* **2014**, *114*, 863–882.

[20] Y. Shirasaki, G. J. Supran, M. G. Bawendi, V. Bulovic. Emergence of colloidal quantum-dot light-emitting technologies. *Nature Photon.* **2013**, *7*, 13–23.

[21] J. O. Winter, T. Y. Liu, B. A. Korgel, C. E. Schmidt. Recognition Molecule Directed Interfacing Between Semiconductor Quantum Dots and Nerve Cells. *Adv. Mater.* **2001**, *13*, 1673–1677.

[22] J. Klostranec, W. Chan. Quantum Dots in Biological and Biomedical Research: Recent Progress and Present Challenges. *Adv. Mater.* **2006**, *18*, 1953–1964.

[23] H. Goesmann, C. Feldmann. Nanoparticulate Functional Materials. *Angew. Chem. Int. Ed.* **2010**, *49*, 1362–1395.

[24] A. P. Alivisatos. Perspectives on the Physical Chemistry of Semiconductor Nanocrystals. *J. Phys. Chem.* **1996**, *100*, 13226–13239.

[25] E. V. Shevchenko, D. V. Talapin, *Self-assembly of semiconductor nanocrystals into ordered superstructures* in *Semiconductor Nanocrystal Quantum Dots*, A. L. Rogach (Ed.), Springer Vienna, **2008**, pp. 119–169.

[26] K. Miszta, J. de Graaf, G. Bertoni, D. Dorfs, R. Brescia, S. Marras, L. Ceseracciu, R. Cingolani, R. van Roij, M. Dijkstra, L. Manna. Hierarchical self-assembly of suspended branched colloidal nanocrystals into superlattice structures. *Nature Mater.* **2011**, *10*, 872–876.

[27] J. Tang, K. W. Kemp, S. Hoogland, K. S. Jeong, H. Liu, L. Levina, M. Furukawa, X. Wang, R. Debnath, D. Cha, K. W. Chou, A. Fischer, A. Amassian, J. B. Asbury, E. H. Sargent. Colloidal-quantum-dot photovoltaics using atomic-ligand passivation. *Nature Mater.* **2011**, *10*, 765–771.

[28] A. H. Ip, S. M. Thon, S. Hoogland, O. Voznyy, D. Zhitomirsky, R. Debnath, L. Levina, L. R. Rollny, G. H. Carey, A. Fischer, K. W. Kemp, I. J. Kramer, Z. Ning, A. J. Labelle, K. W. Chou, A. Amassian, E. H. Sargent. Hybrid passivated colloidal quantum dot solids. *Nature Nanotech.* **2012**, *7*, 577–582.

[29] J. S. Owen, J. Park, P.-E. Trudeau, A. P. Alivisatos. Reaction Chemistry and Ligand Exchange at Cadmium-Selenide Nanocrystal Surfaces. *J. Am. Chem. Soc.* **2008**, *130*, 12279–12281.

[30] W. K. Bae, J. Joo, L. A. Padilha, J. Won, D. C. Lee, Q. Lin, W.-k. Koh, H. Luo, V. I. Klimov, J. M. Pietryga. Highly Effective Surface Passivation of PbSe Quantum Dots through Reaction with Molecular Chlorine. *J. Am. Chem. Soc.* **2012**, *134*, 20160–20168.

Bibliography

[31] N. C. Anderson, J. S. Owen. Soluble, Chloride-Terminated CdSe Nanocrystals: Ligand Exchange Monitored by 1H and 31P NMR Spectroscopy. *Chem. Mater.* **2013**, *25*, 69–76.

[32] M. Zanella, L. Maserati, M. Pernia Leal, M. Prato, R. Lavieville, M. Povia, R. Krahne, L. Manna. Atomic Ligand Passivation of Colloidal Nanocrystal Films via their Reaction with Propyltrichlorosilane. *Chem. Mater.* **2013**, *25*, 1423–1429.

[33] J. Zhang, J. Gao, E. M. Miller, J. M. Luther, M. C. Beard. Diffusion-Controlled Synthesis of PbS and PbSe Quantum Dots with in Situ Halide Passivation for Quantum Dot Solar Cells. *ACS Nano* **2014**, *8*, 614–622.

[34] Y. Xia, Y. Xiong, B. Lim, S. E. Skrabalak. Shape-Controlled Synthesis of Metal Nanocrystals: Simple Chemistry Meets Complex Physics? *Angew. Chem. Int. Ed.* **2009**, *48*, 60–103.

[35] J. S. DuChene, W. Niu, J. M. Abendroth, Q. Sun, W. Zhao, F. Huo, W. D. Wei. Halide Anions as Shape-Directing Agents for Obtaining High-Quality Anisotropic Gold Nanostructures. *Chem. Mater.* **2013**, *25*, 1392–1399.

[36] J. Joo, H. B. Na, T. Yu, J. H. Yu, Y. W. Kim, F. Wu, J. Z. Zhang, T. Hyeon. Generalized and Facile Synthesis of Semiconducting Metal Sulfide Nanocrystals. *J. Am. Chem. Soc.* **2003**, *125*, 11100–11105.

[37] K. An, N. Lee, J. Park, S. C. Kim, Y. Hwang, J.-G. Park, J.-Y. Kim, J.-H. Park, M. J. Han, J. Yu, T. Hyeon. Synthesis, Characterization, and Self-Assembly of Pencil-Shaped CoO Nanorods. *J. Am. Chem. Soc.* **2006**, *128*, 9753–9760.

[38] M. R. Kim, K. Miszta, M. Povia, R. Brescia, S. Christodoulou, M. Prato, S. Marras, L. Manna. Influence of Chloride Ions on the Synthesis of Colloidal Branched CdSe/CdS Nanocrystals by Seeded Growth. *ACS Nano* **2012**, *6*, 11088–11096.

[39] S. Asokan, K. M. Krueger, V. L. Colvin, M. S. Wong. Shape-Controlled Synthesis of CdSe Tetrapods Using Cationic Surfactant Ligands. *Small* **2007**, *3*, 1164–1169.

[40] J. Lim, W. K. Bae, K. U. Park, L. zur Borg, R. Zentel, S. Lee, K. Char. Controlled Synthesis of CdSe Tetrapods with High Morphological Uniformity by the Persistent Kinetic Growth and the Halide-Mediated Phase Transformation. *Chem. Mater.* **2013**, *25*, 1443–1449.

[41] C. Schliehe, B. H. Juárez, M. Pelletier, S. Jander, D. Greshnykh, M. Nagel, A. Meyer, S. Förster, A. Kornowski, C. Klinke, H. Weller. Ultrathin PbS Sheets by Two-Dimensional Oriented Attachment. *Science* **2010**, *329*, 550–553.

[42] C. Schliehe, *Synthese von Bleisulfid Nanostrukturen und Heterosystemen*, Ph.D. thesis, Fachbereich Chemie, Universität Hamburg, **2010**.

[43] V. I. Klimov (Ed.), *Semiconductor and Metal Nanocrystals: Synthesis and Electronic and Optical Properties*, Marcel Dekker, New York, **2004**.

[44] G. Schmid (Ed.), *Nanoparticles: From Theory to Application 2nd ed.*, Wiley-VCH, Weinheim, **2010**.

[45] L. Cademartiri, G. A. Ozin, *Concepts of Nanochemistry*, Wiley VCH, Weinheim, **2009**.

[46] C. d. M. Donega. Synthesis and properties of colloidal heteronanocrystals. *Chem. Soc. Rev.* **2011**, *40*, 1512–1546.

[47] P. Reiss, *Synthesis of semiconductor nanocrystals in organic solvents* in *Semiconductor Nanocrystal Quantum Dots*, A. L. Rogach (Ed.), Springer Vienna, **2008**, pp. 35–72.

[48] M. S. Dresselhaus, G. Dresselhaus, P. Avouris (Eds.), *Carbon Nanotubes: Synthesis, Structure, Properties, and Applications*, Vol. 80 of *Topics in Applied Physics*, Springer, Berlin Heidelberg New York, **2001**.

[49] M. Law, J. Goldberger, P. Yang. Semidconductor Nanowires and Nanotubes. *Annu. Rev. Mater. Res.* **2004**, *34*, 83–122.

[50] H. Li, A. G. Kanaras, L. Manna. Colloidal Branched Semiconductor Nanocrystals: State of the Art and Perspectives. *Acc. Chem. Res.* **2013**, *46*, 1387–1396.

[51] C. Bouet, B. Mahler, B. Nadal, B. Abecassis, M. D. Tessier, S. Ithurria, X. Xu, B. Dubertret. Two-Dimensional Growth of CdSe Nanocrystals, from Nanoplatelets to Nanosheets. *Chem. Mater.* **2013**, *25*, 639–645.

[52] J. Yang, J. S. Son, J. H. Yu, J. Joo, T. Hyeon. Advances in the Colloidal Synthesis of Two-Dimensional Semiconductor Nanoribbons. *Chem. Mater.* **2013**, *25*, 1190–1198.

[53] W. E. Buhro, V. L. Colvin. Semiconductor nanocrystals: Shape matters. *Nature Mater.* **2003**, *2*, 138–139.

[54] S. Polarz. Shape Matters: Anisotropy of the Morphology of Inorganic Colloidal Particles - Synthesis and Function. *Adv. Funct. Mater.* **2011**, *21*, 3214–3230.

[55] A. Salant, M. Shalom, Z. Tachan, S. Buhbut, A. Zaban, U. Banin. Quantum Rod-Sensitized Solar Cell: Nanocrystal Shape Effect on the Photovoltaic Properties. *Nano Lett.* **2012**, *12*, 2095–2100.

[56] A. M. Smith, S. Nie. Semiconductor Nanocrystals: Structure, Properties, and Band Gap Engineering. *Acc. Chem. Res.* **2010**, *43*, 190–200.

[57] L. E. Brus. Electron-electron and electron-hole interactions in small semiconductor crystallites: The size dependence of the lowest excited electronic state. *J. Chem. Phys.* **1984**, *80*, 4403–4409.

[58] L. Brus. Electronic wave functions in semiconductor clusters: experiment and theory. *J. Phys. Chem.* **1986**, *90*, 2555–2560.

[59] V. V. Sobolev, V. I. Donetskina, E. F. Zagainov. Direct precision method for detection of excitons in II-VI and crystals at room and liquid nitrogen temperatures. *Sov. Phys. Semicond.* **1978**, *12*, 646. *Fiz Tekh. Poluprovodn.* **1978**, *12*, 1089.

[60] H. E. Swanson, N. T. Gilfrich, M. I. Cook. Standard X-ray diffraction powder patterns. *Natl. Bur. Stand. (U.S), Circ. 539* **1957**, *7*, 12; JCPDS # 00–008–0459.

[61] C.-Y. Yeh, Z. W. Lu, S. Froyen, A. Zunger. Zinc-blende-wurtzite polytypism in semiconductors. *Phys. Rev. B* **1992**, *46*, 10086–10097.

[62] G. Tai, J. Zhou, W. Guo. Inorganic salt-induced phase control and optical characterization of cadmium sulfide nanoparticles. *Nanotechnology* **2010**, *21*, 175601.

[63] T.-T. Zhuang, P. Yu, F.-J. Fan, L. Wu, X.-J. Liu, S.-H. Yu. Controlled Synthesis of Kinked Ultrathin ZnS Nanorods/Nanowires Triggered by Chloride Ions: A Case Study. *Small* **2013**, *10*, 1394–1402.

[64] J. J. Shiang, A. V. Kadavanich, R. K. Grubbs, A. P. Alivisatos. Symmetry of Annealed Wurtzite CdSe Nanocrystals: Assignment to the C3v Point Group. *J. Phys. Chem.* **1995**, *99*, 17417–17422.

[65] G. A. Wolff, J. J. Frawley, J. Hietanen. On the Etching of II-VI and III-V Compounds. *J. Electrochem. Soc.* **1964**, *111*, 22–27.

[66] G. A. Wolff, J. D. Broder. The role of ionicity, bonding and adsorption in crystal morphology. *Colloq Int CNRS* **1965**, *152*, 171–92.

[67] S. A. Blanton, R. L. Leheny, M. A. Hines, P. Guyot-Sionnest. Dielectric Dispersion Measurements of CdSe Nanocrystal Colloids: Observation of a Permanent Dipole Moment. *Phys. Rev. Lett.* **1997**, *79*, 865–868.

[68] M. Shim, P. Guyot-Sionnest. Permanent dipole moment and charges in colloidal semiconductor quantum dots. *J. Chem. Phys.* **1999**, *111*, 6955–6964.

[69] J. Jasieniak, M. Califano, S. E. Watkins. Size-Dependent Valence and Conduction Band-Edge Energies of Semiconductor Nanocrystals. *ACS Nano* **2011**, *5*, 5888–5902.

[70] P. Guyot-Sionnest. Electrical Transport in Colloidal Quantum Dot Films. *J. Phys. Chem. Lett.* **2012**, *3*, 1169–1175.

[71] X. Peng. Mechanisms for the Shape-Control and Shape-Evolution of Colloidal Semiconductor Nanocrystals. *Adv. Mater.* **2003**, *15*, 459–463.

[72] L. Cademartiri, G. A. Ozin, *Concepts of Nanochemistry*, Wiley VCH, Weinheim, **2009**, pp. 153-154.

[73] W. Ostwald. Über die vermeintliche Isomerie des roten und gelben Quecksilberoxyds und die Oberflächenspannung fester Körper. *Z. Phys. Chem.* **1900**, *34*, 495–503.

[74] S.-M. Lee, S.-N. Cho, J. Cheon. Anisotropic Shape Control of Colloidal Inorganic Nanocrystals. *Adv. Mater.* **2003**, *15*, 441–444.

[75] Y. Yin, A. P. Alivisatos. Colloidal nanocrystal synthesis and the organic-inorganic interface. *Nature* **2005**, *437*, 664–670.

[76] S. Kumar, T. Nann. Shape Control of II-VI Semiconductor Nanomaterials. *Small* **2006**, *2*, 316–329.

[77] L. Manna, L. Wang, R. Cingolani, A. P. Alivisatos. First-Principles Modeling of Unpassivated and Surfactant-Passivated Bulk Facets of Wurtzite CdSe: A Model System for Studying the Anisotropic Growth of CdSe Nanocrystals. *J. Chem. Phys. B* **2005**, *109*, 6183–6192.

[78] A. J. Morris-Cohen, M. D. Donakowski, K. E. Knowles, E. A. Weiss. The Effect of a Common Purification Procedure on the Chemical Composition of the Surfaces of CdSe Quantum Dots Synthesized with Trioctylphosphine Oxide. *J. Chem. Phys. C* **2010**, *114*, 897–906.

[79] X. Peng, J. Wickham, A. P. Alivisatos. Kinetics of II-VI and III-V Colloidal Semiconductor Nanocrystal Growth: Focusing of Size Distributions. *J. Am. Chem. Soc.* **1998**, *120*, 5343–5344.

[80] J. W. Mullin, *Crystallization 3rd ed.*, Butterworth-Heinemann, Oxford, **1997**.

[81] I. Sunagawa, *Crystals: Growth, Morphology and Perfection*, Cambrige University Press, Cambridge, **2005**.

[82] L. Manna, E. C. Scher, A. P. Alivisatos. Synthesis of Soluble and Processable Rod-, Arrow-, Teardrop-, and Tetrapod-Shaped CdSe Nanocrystals. *J. Am. Chem. Soc.* **2000**, *122*, 12700–12706.

[83] A. P. Alivisatos. Naturally Aligned Nanocrystals. *Science* **2000**, *289*, 736–737.

[84] Q. Zhang, S.-J. Liu, S.-H. Yu. Recent advances in oriented attachment growth and synthesis of functional materials: concept, evidence, mechanism, and future. *J. Mater. Chem.* **2009**, *19*, 191–207.

[85] W.-k. Koh, A. C. Bartnik, F. W. Wise, C. B. Murray. Synthesis of Monodisperse PbSe Nanorods: A Case for Oriented Attachment. *J. Am. Chem. Soc.* **2010**, *132*, 3909–3913.

[86] K.-S. Cho, D. V. Talapin, W. Gaschler, C. B. Murray. Designing PbSe Nanowires and Nanorings through Oriented Attachment of Nanoparticles. *J. Am. Chem. Soc.* **2005**, *127*, 7140–7147.

[87] E. Lifshitz, M. Bashouti, V. Kloper, A. Kigel, M. S. Eisen, S. Berger. Synthesis and Characterization of PbSe Quantum Wires, Multipods, Quantum Rods, and Cubes. *Nano Lett.* **2003**, *3*, 857–862.

[88] Y.-H. Liu, F. Wang, Y. Wang, P. C. Gibbons, W. E. Buhro. Lamellar Assembly of Cadmium Selenide Nanoclusters into Quantum Belts. *J. Am. Chem. Soc.* **2011**, *133*, 17005–17013.

[89] X. Peng, L. Manna, W. Yang, J. Wickham, E. Scher, A. Kadavanich, A. P. Alivisatos. Shape control of CdSe nanocrystals. *Nature* **2000**, *404*, 59–61.

[90] Z. A. Peng, X. Peng. Mechanisms of the Shape Evolution of CdSe Nanocrystals. *J. Am. Chem. Soc.* **2001**, *123*, 1389–1395.

[91] Z. A. Peng, X. Peng. Formation of High-Quality CdTe, CdSe, and CdS Nanocrystals Using CdO as Precursor. *J. Am. Chem. Soc.* **2001**, *123*, 183–184.

[92] Z. A. Peng, X. Peng. Nearly Monodisperse and Shape-Controlled CdSe Nanocrystals via Alternative Routes: Nucleation and Growth. *J. Am. Chem. Soc.* **2002**, *124*, 3343–3353.

[93] V. K. LaMer, R. H. Dinegar. Theory, Production and Mechanism of Formation of Monodispersed Hydrosols. *J. Am. Chem. Soc.* **1950**, *72*, 4847–4854.

[94] S. Abe, R. K. Čapek, B. De Geyter, Z. Hens. Reaction Chemistry/Nanocrystal Property Relations in the Hot Injection Synthesis, the Role of the Solute Solubility. *ACS Nano* **2013**, *7*, 943–949.

[95] R. García-Rodríguez, M. P. Hendricks, B. M. Cossairt, H. Liu, J. S. Owen. Conversion Reactions of Cadmium Chalcogenide Nanocrystal Precursors. *Chem. Mater.* **2013**, *25*, 1233–1249.

[96] A. Puzder, A. J. Williamson, N. Zaitseva, G. Galli, L. Manna, A. P. Alivisatos. The Effect of Organic Ligand Binding on the Growth of CdSe Nanoparticles Probed by Ab Initio Calculations. *Nano Lett.* **2004**, *4*, 2361–2365.

[97] J. Y. Rempel, B. L. Trout, M. G. Bawendi, K. F. Jensen. Density Functional Theory Study of Ligand Binding on CdSe (0001), (000$\bar{1}$), and (11$\bar{2}$0) Single Crystal Relaxed and Reconstructed Surfaces: Implications for Nanocrystalline Growth. *J. Phys. Chem. B* **2006**, *110*, 18007–18016.

[98] S. M. Hughes, A. P. Alivisatos. Anisotropic Formation and Distribution of Stacking Faults in II-VI Semiconductor Nanorods. *Nano Lett.* **2013**, *13*, 106–110.

[99] M. D. Clark, S. K. Kumar, J. S. Owen, E. M. Chan. Focusing Nanocrystal Size Distributions via Production Control. *Nano Lett.* **2011**, *11*, 1976–1980.

[100] L. Carbone, S. Kudera, E. Carlino, W. J. Parak, C. Giannini, R. Cingolani, L. Manna. Multiple Wurtzite Twinning in CdTe Nanocrystals Induced by Methylphosphonic Acid. *J. Am. Chem. Soc.* **2006**, *128*, 748–755.

[101] A. Wolcott, R. C. Fitzmorris, O. Muzaffery, J. Z. Zhang. CdSe Quantum Rod Formation Aided By In Situ TOPO Oxidation. *Chem. Mater.* **2010**, *22*, 2814–2821.

[102] F. Wang, R. Tang, W. E. Buhro. The Trouble with TOPO; Identification of Adventitious Impurities Beneficial to the Growth of Cadmium Selenide Quantum Dots, Rods, and Wires. *Nano Lett.* **2008**, *8*, 3521–3524.

[103] F. Wang, R. Tang, J. L.-F. Kao, S. D. Dingman, W. E. Buhro. Spectroscopic Identification of Tri-n-octylphosphine Oxide (TOPO) Impurities and Elucidation of Their Roles in Cadmium Selenide Quantum-Wire Growth. *J. Am. Chem. Soc.* **2009**, *131*, 4983–4994.

[104] C. M. Evans, M. E. Evans, T. D. Krauss. Mysteries of TOPSe Revealed: Insights into Quantum Dot Nucleation. *J. Am. Chem. Soc.* **2010**, *132*, 10973–10975.

[105] M. Meyns, N. G. Bastús, Y. Cai, A. Kornowski, B. H. Juárez, H. Weller, C. Klinke. Growth and reductive transformation of a gold shell around pyramidal cadmium selenide nanocrystals. *J. Mater. Chem.* **2010**, *20*, 10602–10605.

[106] A. Hungría, B. Juárez, C. Klinke, H. Weller, P. Midgley. 3-D characterization of CdSe nanoparticles attached to carbon nanotubes. *Nano Res.* **2008**, *1*, 89–97.

[107] S. Abe, R. K. Čapek, B. De Geyter, Z. Hens. Tuning the Postfocused Size of Colloidal Nanocrystals by the Reaction Rate: From Theory to Application. *ACS Nano* **2012**, *6*, 42–53.

[108] N. C. Anderson, M. P. Hendricks, J. J. Choi, J. S. Owen. Ligand Exchange and the Stoichiometry of Metal Chalcogenide Nanocrystals: Spectroscopic Observation of Facile Metal-Carboxylate Displacement and Binding. *J. Am. Chem. Soc.* **2013**, *135*, 18536–18548.

[109] X. De-Bao, H. Ztt-Yum, X. Lu, Z. Li-Yim, Y. Wen-Sheng, Y. Jian-Nian. Effect of Particle Size and Capping on Photoluminescence Quantum Efficiency of 1,3,5-Triphenyl-2-pyrazoline Nanocrystals. *Chin. J. Chem.* **2003**, *21*, 79–82.

[110] J. Geweke, *Synthese pyramidaler CdSe-Nanopartikel und der Einfluss chlorhaltiger Additive auf ihre Form.*, Bachelor thesis, Fachbereich Chemie, Universität Hamburg, **2011**.

[111] L. Qu, Z. A. Peng, X. Peng. Alternative Routes toward High Quality CdSe Nanocrystals. *Nano Lett.* **2001**, *1*, 333–337.

[112] W. W. Yu, X. Peng. Formation of High-Quality CdS and Other II-VI Semiconductor Nanocrystals in Noncoordinating Solvents: Tunable Reactivity of Monomers. *Angew. Chem. Int. Ed.* **2002**, *41*, 2368–2371.

[113] C. R. Bullen, P. Mulvaney. Nucleation and Growth Kinetics of CdSe Nanocrystals in Octadecene. *Nano Lett.* **2004**, *4*, 2303–2307.

[114] J. van Embden, P. Mulvaney. Nucleation and Growth of CdSe Nanocrystals in a Binary Ligand System. *Langmuir* **2005**, *21*, 10226–10233.

[115] G. G. Yordanov, H. Yoshimura, C. D. Dushkin. Fine control of the growth and optical properties of CdSe quantum dots by varying the amount of stearic acid in a liquid paraffin matrix. *Colloids Surf., A* **2008**, *322*, 177–182.

[116] J. S. Owen, E. M. Chan, H. Liu, A. P. Alivisatos. Precursor Conversion Kinetics and the Nucleation of Cadmium Selenide Nanocrystals. *J. Am. Chem. Soc.* **2010**, *132*, 18206–18213.

[117] Y. Zou, D. Li, D. Yang. Shape and phase control of CdS nanocrystals using cationic surfactant in noninjection synthesis. *Nanoscale Res. Lett.* **2011**, *6*, 1–6.

[118] F. Wang, W. E. Buhro. Morphology Control of Cadmium Selenide Nanocrystals: Insights into the Roles of Di-n-octylphosphine Oxide (DOPO) and Di-n-octylphosphinic Acid (DOPA). *J. Am. Chem. Soc.* **2012**, *134*, 5369–5380.

[119] L. Manna, D. J. Milliron, A. Meisel, E. C. Scher, A. P. Alivisatos. Controlled growth of tetrapod-branched inorganic nanocrystals. *Nature Mater.* **2003**, *2*, 382–385.

[120] B. Böhme, S. Hoffmann, M. Baitinger, Y. Grin. Application of n-Dodecyltrimethylammonium Chloride for the Oxidation of Intermetallic Phases. *Z. Naturforsch. B* **2011**, *66*, 230–238.

[121] M. Bohnet, F. Ullmann, *Ullmann's Encyclopedia of Industrial Chemistry 6th ed.*, *Vol. 8: Chlorinated hydrocarbons to Cobalt and cobalt compounds*, Wiley VCH, Weinheim, **2003**, pp. 34-35.

[122] A. W. Heijden, A. J. Mens, R. Bogerd, B. M. Weckhuysen. Dehydrochlorination of Intermediates in the Production of Vinyl Chloride over Lanthanum Oxide-Based Catalysts. *Catal. Lett.* **2008**, *122*, 238–246.

[123] C. Palencia, K. Lauwaet, L. de la Cueva, M. Acebrón, J. J. Conde, M. Meyns, C. Klinke, J. M. Gallego, R. Otero, B. H. Juárez. Cl-capped CdSe nanocrystals via in situ generation of chloride anions. *Nanoscale* **2014**, *6*, 6812–6818.

[124] M. Meyns, F. Iacono, C. Palencia, J. Geweke, M. D. Coderch, U. E. A. Fittschen, J. M. Gallego, R. Otero, B. H. Juárez, C. Klinke. Shape Evolution of CdSe Nanoparticles Controlled by Halogen Compounds. *Chem. Mater.* **2014**, *26*, 1813–1821.

[125] F. Iacono, C. Palencia, L. de la Cueva, M. Meyns, L. Terracciano, A. Vollmer, M. de la Mata, C. Klinke, J. M. Gallego, B. H. Juárez, R. Otero. Interfacing Quantum Dots and Graphitic Surfaces with Chlorine Atomic Ligands. *ACS Nano* **2013**, *7*, 2559–2565.

[126] R. D. Seals, R. Alexander, L. T. Taylor, J. G. Dillard. Core electron binding energy study of group IIb-VIIa compounds. *Inorg. Chem.* **1973**, *12*, 2485–2487.

[127] J. Taylor, T. Kippeny, S. J. Rosenthal. Surface Stoichiometry of CdSe Nanocrystals Determined by Rutherford Backscattering Spectroscopy. *J. Cluster Sci.* **2001**, *12*, 571–582.

[128] J. W. Anthony, R. A. Bideaux, K. W. Bladh, E. Nichols, M. C., *Handbook of Mineralogy*, [Online], **2013**, Last accessed 22 July **2014**. http://www.handbookofmineralogy.org/.

[129] F. Neese. The ORCA program system. *WIREs Comput Mol Sci* **2012**, *2*, 73–78.

[130] S. H. Vosko, L. Wilk, M. Nusair. Accurate spin-dependent electron liquid correlation energies for local spin density calculations: a critical analysis. *Can. J. Phys.* **1980**, *58*, 1200–1211.

[131] A. Schafer, H. Horn, R. Ahlrichs. Fully optimized contracted Gaussian basis sets for atoms Li to Kr. *J. Chem. Phys.* **1992**, *97*, 2571–2577.

[132] Y. Cai, *Hochgeordnete 2D Nanopartikelfilme: Herstellung, Charakterisierung und elektrische Transportmessungen*, Ph.D. thesis, Fachbereich Chemie, Universität Hamburg, **2012**.

[133] E. Rabani. Structure and electrostatic properties of passivated CdSe nanocrystals. *J. Chem. Phys.* **2001**, *115*, 1493–1497.

[134] R. Li, J. Lee, B. Yang, D. N. Horspool, M. Aindow, F. Papadimitrakopoulos. Amine-Assisted Facetted Etching of CdSe Nanocrystals. *J. Am. Chem. Soc.* **2005**, *127*, 2524–2532.

[135] S. J. Lim, W. Kim, S. Jung, J. Seo, S. K. Shin. Anisotropic Etching of Semiconductor Nanocrystals. *Chem. Mater.* **2011**, *23*, 5029–5036.

[136] S. J. Lim, W. Kim, S. K. Shin. Surface-Dependent, Ligand-Mediated Photochemical Etching of CdSe Nanoplatelets. *J. Am. Chem. Soc.* **2012**, *134*, 7576–7579.

[137] M. Lamoureux, J. Milne. Selenium chloride and bromide equilibria in aprotic solvents; a ^{77}Se NMR study. *Polyhedron* **1990**, *9*, 589–595.

[138] H. Liu, J. S. Owen, A. P. Alivisatos. Mechanistic Study of Precursor Evolution in Colloidal Group II-VI Semiconductor Nanocrystal Synthesis. *J. Am. Chem. Soc.* **2007**, *129*, 305–312.

[139] Z.-J. Jiang, D. F. Kelley. Role of Magic-Sized Clusters in the Synthesis of CdSe Nanorods. *ACS Nano* **2010**, *4*, 1561–1572.

[140] M. Saruyama, M. Kanehara, T. Teranishi. Drastic Structural Transformation of Cadmium Chalcogenide Nanoparticles Using Chloride Ions and Surfactants. *J. Am. Chem. Soc.* **2010**, *132*, 3280–3282.

[141] A. Kudo, Y. Miseki. Heterogeneous photocatalyst materials for water splitting. *Chem. Soc. Rev.* **2009**, *38*, 253–278.

[142] S. Rawalekar, T. Mokari. Rational Design of Hybrid Nanostructures for Advanced Photocatalysis. *Adv. Energy Mater.* **2013**, *3*, 12–27.

[143] K. Liu, M. Sakurai, M. Liao, M. Aono. Giant Improvement of the Performance of ZnO Nanowire Photodetectors by Au Nanoparticles. *J. Phys. Chem. C* **2010**, *114*, 19835–19839.

[144] T. Pellegrino, S. Kudera, T. Liedl, A. Muñoz Javier, L. Manna, W. J. Parak. On the Development of Colloidal Nanoparticles towards Multifunctional Structures and their Possible Use for Biological Applications. *Small* **2005**, *1*, 48–63.

[145] X. Liu, C. Lee, W.-C. Law, D. Zhu, M. Liu, M. Jeon, J. Kim, P. N. Prasad, C. Kim, M. T. Swihart. Au-$Cu_{2-x}Se$ Heterodimer Nanoparticles with Broad Localized Surface Plasmon Resonance as Contrast Agents for Deep Tissue Imaging. *Nano Lett.* **2013**, *13*, 4333–4339.

[146] M. R. Buck, R. E. Schaak. Emerging Strategies for the Total Synthesis of Inorganic Nanostructures. *Angew. Chem. Int. Ed.* **2013**, *52*, 6154–6178.

[147] M. Ibáñez, A. Cabot. All Change for Nanocrystals. *Science* **2013**, *340*, 935–936.

[148] S. Gupta, S. V. Kershaw, A. L. Rogach. 25th Anniversary Article: Ion Exchange in Colloidal Nanocrystals. *Adv. Mater.* **2013**, *25*, 6923–6944.

[149] E. Shaviv, U. Banin. Synergistic Effects on Second Harmonic Generation of Hybrid CdSe-Au Nanoparticles. *ACS Nano* **2010**, *4*, 1529–1538.

[150] H. Kim, M. Achermann, L. P. Balet, J. A. Hollingsworth, V. I. Klimov. Synthesis and Characterization of Co/CdSe Core/Shell Nanocomposites: Bifunctional Magnetic-Optical Nanocrystals. *J. Am. Chem. Soc.* **2005**, *127*, 544–546.

[151] A. Vaneski, A. S. Susha, J. Rodríguez-Fernández, M. Berr, F. Jäckel, J. Feldmann, A. L. Rogach. Hybrid Colloidal Heterostructures of Anisotropic Semiconductor Nanocrystals Decorated with Noble Metals: Synthesis and Function. *Adv. Funct. Mater.* **2011**, *21*, 1547–1556.

[152] U. Banin, Y. Ben-Shahar, K. Vinokurov. Hybrid Semiconductor-Metal Nanoparticles: From Architecture to Function. *Chem. Mater.* **2014**, *26*, 97–110.

[153] M. Casavola, R. Buonsanti, G. Caputo, P. D. Cozzoli. Colloidal Strategies for Preparing Oxide-Based Hybrid Nanocrystals. *Eur. J. Inorg. Chem.* **2008**, 837–854.

[154] L. Carbone, P. D. Cozzoli. Colloidal heterostructured nanocrystals: Synthesis and growth mechanisms. *Nano Today* **2010**, *5*, 449–493.

[155] P. Kundu, E. A. Anumol, C. Nethravathi, N. Ravishankar. Existing and emerging strategies for the synthesis of nanoscale heterostructures. *Phys. Chem. Chem. Phys.* **2011**, *13*, 19256–19269.

[156] P. V. Kamat. Meeting the Clean Energy Demand: Nanostructure Architectures for Solar Energy Conversion. *J. Phys. Chem. C* **2007**, *111*, 2834–2860.

[157] A. Figuerola, I. R. Franchini, A. Fiore, R. Mastria, A. Falqui, G. Bertoni, S. Bals, G. Van Tendeloo, S. Kudera, R. Cingolani, L. Manna. End-to-End Assembly of Shape-Controlled Nanocrystals via a Nanowelding Approach Mediated by Gold Domains. *Adv. Mater.* **2009**, *21*, 550–554.

[158] R. Lavieville, Y. Zhang, A. Casu, A. Genovese, L. Manna, E. Di Fabrizio, R. Krahne. Charge Transport in Nanoscale "All-Inorganic" Networks of Semiconductor Nanorods Linked by Metal Domains. *ACS Nano* **2012**, *6*, 2940–2947.

[159] W. Shi, Y. Sahoo, H. Zeng, Y. Ding, M. T. Swihart, P. N. Prasad. Anisotropic Growth of PbSe Nanocrystals on Au-Fe_3O_4 Hybrid Nanoparticles. *Adv. Mater.* **2006**, *18*, 1889–1894.

[160] A. E. Saunders, I. Popov, U. Banin. Synthesis of Hybrid CdS-Au Colloidal Nanostructures. *J. Phys. Chem. B* **2006**, *110*, 25421–25429.

[161] G. Menagen, J. E. Macdonald, Y. Shemesh, I. Popov, U. Banin. Au Growth on Semiconductor Nanorods: Photoinduced versus Thermal Growth Mechanisms. *J. Am. Chem. Soc.* **2009**, *131*, 17406–17411.

[162] T. Bala, A. Sanyal, A. Singh, D. Kelly, C. O'Sullivan, F. Laffir, K. M. Ryan. Silver tip formation on colloidal CdSe nanorods by a facile phase transfer protocol. *J. Mater. Chem.* **2011**, *21*, 6815–6820.

[163] S. E. Habas, P. Yang, T. Mokari. Selective Growth of Metal and Binary Metal Tips on CdS Nanorods. *J. Am. Chem. Soc.* **2008**, *130*, 3294–3295.

[164] H. Schlicke, D. Ghosh, L.-K. Fong, H. L. Xin, H. Zheng, A. P. Alivisatos. Selective Placement of Faceted Metal Tips on Semiconductor Nanorods. *Angew. Chem. Int. Ed.* **2013**, *52*, 980–982.

Bibliography

[165] J.-S. Lee, E. V. Shevchenko, D. V. Talapin. Au-PbS Core-Shell Nanocrystals: Plasmonic Absorption Enhancement and Electrical Doping via Intra-particle Charge Transfer. *J. Am. Chem. Soc.* **2008**, *130*, 9673–9675.

[166] J. Zhang, Y. Tang, K. Lee, M. Ouyang. Nonepitaxial Growth of Hybrid Core-Shell Nanostructures with Large Lattice Mismatches. *Science* **2010**, *327*, 1634–1638.

[167] T. Pellegrino, A. Fiore, E. Carlino, C. Giannini, P. D. Cozzoli, G. Ciccarella, M. Respaud, L. Palmirotta, R. Cingolani, L. Manna. Heterodimers Based on $CoPt_3$-Au Nanocrystals with Tunable Domain Size. *J. Am. Chem. Soc.* **2006**, *128*, 6690–6698.

[168] J. Zeng, J. Huang, C. Liu, C. H. Wu, Y. Lin, X. Wang, S. Zhang, J. Hou, Y. Xia. Gold-Based Hybrid Nanocrystals Through Heterogeneous Nucleation and Growth. *Adv. Mater.* **2010**, *22*, 1936–1940.

[169] A. Figuerola, M. v. Huis, M. Zanella, A. Genovese, S. Marras, A. Falqui, H. W. Zandbergen, R. Cingolani, L. Manna. Epitaxial CdSe-Au Nanocrystal Heterostructures by Thermal Annealing. *Nano Lett.* **2010**, *10*, 3028–3036.

[170] H.-B. Yao, Y. Guan, J. Zheng, G. Huang, J. Xu, J.-W. Liu, H.-P. Cong, S.-H. Yu. Chloride Anion Triggered Synthesis and Assembly of Gold Nanoparticle-Ultrathin Cadmium Selenide Nanowire Networks with Enhanced Photoconductivity. *Part. Part. Sys. Char.* **2013**, *30*, 97–101.

[171] J. Yang, H. I. Elim, Q. Zhang, J. Y. Lee, W. Ji. Rational Synthesis, Self-Assembly, and Optical Properties of PbS-Au Heterogeneous Nanostructures via Preferential Deposition. *J. Am. Chem. Soc.* **2006**, *128*, 11921–11926.

[172] S. Huang, J. Huang, J. Yang, J.-J. Peng, Q. Zhang, F. Peng, H. Wang, H. Yu. Chemical Synthesis, Structure Characterization, and Optical Properties of Hollow PbS_x-Solid Au Heterodimer Nanostructures. *Chem. Eur. J.* **2010**, *16*, 5920–5926.

[173] T. Mokari, R. Costi, C. G. Sztrum, E. Rabani, U. Banin. Formation of symmetric and asymmetric metal-semiconductor hybrid nanoparticles. *Phys. Status Solidi B* **2006**, *243*, 3952–3958.

[174] I. Jen-La Plante, S. E. Habas, B. D. Yuhas, D. J. Gargas, T. Mokari. Interfacing Metal Nanoparticles with Semiconductor Nanowires. *Chem. Mater.* **2009**, *21*, 3662–3667.

[175] L. L. Chng, J. Yang, Y. Wei, J. Y. Ying. Semiconductor-Gold Nanocomposite Catalysts for the Efficient Three-Component Coupling of Aldehyde, Amine and Alkyne in Water. *Adv. Synth. Catal.* **2009**, *351*, 2887–2896.

[176] F. Jin, M.-L. Zhang, M.-L. Zheng, Z.-H. Liu, Y.-M. Fan, K. Xu, Z.-S. Zhao, X.-M. Duan. Hierarchical CdSe-gold hybrid nanocrystals: synthesis and optical properties. *Phys. Chem. Chem. Phys.* **2012**, *14*, 13180–13186.

[177] T. Mokari. Synthesis and characterization of hybrid nanostructures. *Nano Reviews* **2011**, *2*.

[178] N. Mishra, J. Lian, S. Chakrabortty, M. Lin, Y. Chan. Unusual Selectivity of Metal Deposition on Tapered Semiconductor Nanostructures. *Chem. Mater.* **2012**, *24*, 2040–2046.

[179] Z. Zhang, J. T. Yates. Band Bending in Semiconductors: Chemical and Physical Consequences at Surfaces and Interfaces. *Chem. Rev.* **2012**, *112*, 5520–5551.

[180] E. Khon, A. Mereshchenko, A. N. Tarnovsky, K. Acharya, A. Klinkova, N. N. Hewa-Kasakarage, I. Nemitz, M. Zamkov. Suppression of the Plasmon Resonance in Au/CdS Colloidal Nanocomposites. *Nano Lett.* **2011**, *11*, 1792–1799.

[181] R. K. Swank. Surface Properties of II-VI Compounds. *Phys. Rev.* **1967**, *153*, 844–849.

[182] D. R. Lide (Ed.), *CRC Handbook of Chemistry and Physics 89th ed.*, CRC Press, Boca Raton, FL, **2008**, p. 12-114.

[183] D. M. Alloway, M. Hofmann, D. L. Smith, N. E. Gruhn, A. L. Graham, R. Colorado, V. H. Wysocki, T. R. Lee, P. A. Lee, N. R. Armstrong. Interface Dipoles Arising from Self-Assembled Monolayers on Gold: UV-Photoemission Studies of Alkanethiols and Partially Fluorinated Alkanethiols. *J. Phys. Chem. B* **2003**, *107*, 11690–11699.

[184] J. Jasieniak, L. Smith, J. v. Embden, P. Mulvaney, M. Califano. Re-examination of the Size-Dependent Absorption Properties of CdSe Quantum Dots. *J. Phys. Chem. C* **2009**, *113*, 19468–19474.

[185] R. G. Wheeler, J. O. Dimmock. Exciton Structure and Zeeman Effects in Cadmium Selenide. *Phys. Rev.* **1962**, *125*, 1805–1815.

[186] S. Linic, P. Christopher, D. B. Ingram. Plasmonic-metal nanostructures for efficient conversion of solar to chemical energy. *Nature Mater.* **2011**, *10*, 911–921.

[187] M. Achermann. Exciton-Plasmon Interactions in Metal-Semiconductor Nanostructures. *J. Phys. Chem. Lett.* **2010**, *1*, 2837–2843.

[188] J. Zhang, Y. Tang, K. Lee, M. Ouyang. Tailoring light-matter-spin interactions in colloidal hetero-nanostructures. *Nature* **2010**, *466*, 91–95.

[189] S. Link, M. A. El-Sayed. Shape and size dependence of radiative, non-radiative and photothermal properties of gold nanocrystals. *Int. Rev. Phys. Chem.* **2000**, *19*, 409–453.

[190] G. Schmid, *Nanoscale Materials in Chemistry* in *Nanoscale Materials in Chemistry*, K. J. Klabunde (Ed.), John Wiley & Sons, Inc., **2002**, chapter Metals, pp. 15–59.

[191] L. M. Liz-Marzán. Nanometals: Formation and color. *Mater. Today* **2004**, *7*, 26–31.

[192] M. Grzelczak, L. M. Liz-Marzán. Colloidal Nanoplasmonics: From Building Blocks to Sensing Devices. *Langmuir* **2013**, *29*, 4652–4663.

[193] S. Malola, L. Lehtovaara, J. Enkovaara, H. Häkkinen. Birth of the Localized Surface Plasmon Resonance in Monolayer-Protected Gold Nanoclusters. *ACS Nano* **2013**, *7*, 10263–10270.

[194] J. A. Creighton, D. G. Eadon. Ultraviolet-visible absorption spectra of the colloidal metallic elements. *J. Chem. Soc., Faraday Trans.* **1991**, *87*, 3881–3891.

[195] O. Kulakovich, N. Strekal, A. Yaroshevich, S. Maskevich, S. Gaponenko, I. Nabiev, U. Woggon, M. Artemyev. Enhanced Luminescence of CdSe Quantum Dots on Gold Colloids. *Nano Lett.* **2002**, *2*, 1449–1452.

[196] Y. Fedutik, V. Temnov, U. Woggon, E. Ustinovich, M. Artemyev. Exciton-Plasmon Interaction in a Composite Metal-Insulator-Semiconductor Nanowire System. *J. Am. Chem. Soc.* **2007**, *129*, 14939–14945.

[197] B. P. Khanal, A. Pandey, L. Li, Q. Lin, W. K. Bae, H. Luo, V. I. Klimov, J. M. Pietryga. Generalized Synthesis of Hybrid Metal-Semiconductor Nanostructures Tunable from the Visible to the Infrared. *ACS Nano* **2012**, *6*, 3832–3840.

[198] P. V. Kamat, B. Shanghavi. Interparticle Electron Transfer in Metal/Semiconductor Composites. Picosecond Dynamics of CdS-Capped Gold Nanoclusters. *J. Phys. Chem. B* **1997**, *101*, 7675–7679.

[199] P. Yu, X. Wen, Y.-C. Lee, W.-C. Lee, C.-C. Kang, J. Tang. Photoinduced Ultrafast Charge Separation in Plexcitonic CdSe/Au and CdSe/Pt Nanorods. *J. Phys. Chem. Lett.* **2013**, *4*, 3596–3601.

[200] L. Carbone, A. Jakab, Y. Khalavka, C. Sönnichsen. Light-Controlled One-Sided Growth of Large Plasmonic Gold Domains on Quantum Rods Observed on the Single Particle Level. *Nano Lett.* **2009**, *9*, 3710–3714.

[201] A. O. Govorov, G. W. Bryant, W. Zhang, T. Skeini, J. Lee, N. A. Kotov, J. M. Slocik, R. R. Naik. Exciton-Plasmon Interaction and Hybrid Excitons in Semiconductor-Metal Nanoparticle Assemblies. *Nano Lett.* **2006**, *6*, 984–994.

[202] T. Shanmugapriya, P. Ramamurthy. Photoluminescence Enhancement of Nanogold Decorated CdS Quantum Dots. *J. Phys. Chem. C* **2013**, *117*, 12272–12278.

[203] F. Sastre, M. Oteri, A. Corma, H. Garcia. Photocatalytic water gas shift using visible or simulated solar light for the efficient, room-temperature hydrogen generation. *Energy Environ. Sci.* **2013**, *6*, 2211–2215.

[204] R. Costi, G. Cohen, A. Salant, E. Rabani, U. Banin. Electrostatic Force Microscopy Study of Single Au-CdSe Hybrid Nanodumbbells: Evidence for Light-Induced Charge Separation. *Nano Lett.* **2009**, *9*, 2031–2039.

[205] D. Mongin, E. Shaviv, P. Maioli, A. Crut, U. Banin, N. Del Fatti, F. Vallée. Ultrafast Photoinduced Charge Separation in Metal-Semiconductor Nanohybrids. *ACS Nano* **2012**, *6*, 7034–7043.

[206] A. Wood, M. Giersig, P. Mulvaney. Fermi Level Equilibration in Quantum Dot-Metal Nanojunctions. *J. Phys. Chem. B* **2001**, *105*, 8810–8815.

[207] V. Subramanian, E. E. Wolf, P. V. Kamat. Green Emission to Probe Photoinduced Charging Events in ZnO-Au Nanoparticles. Charge Distribution and Fermi-Level Equilibration. *J. Phys. Chem. B* **2003**, *107*, 7479–7485.

[208] P. D. Cozzoli, M. L. Curri, A. Agostiano. Efficient charge storage in photoexcited TiO_2 nanorod-noble metal nanoparticle composite systems. *Chem. Commun.* **2005**, 3186–3188.

[209] J. U. Bang, S. J. Lee, J. S. Jang, W. Choi, H. Song. Geometric Effect of Single or Double Metal-Tipped CdSe Nanorods on Photocatalytic H_2 Generation. *J. Phys. Chem. Lett.* **2012**, *3*, 3781–3785.

[210] E. Elmalem, A. E. Saunders, R. Costi, A. Salant, U. Banin. Growth of Photocatalytic CdSe-Pt Nanorods and Nanonets. *Adv. Mater.* **2008**, *20*, 4312–4317.

[211] N. Bao, L. Shen, T. Takata, K. Domen. Self-Templated Synthesis of Nanoporous CdS Nanostructures for Highly Efficient Photocatalytic Hydrogen Production under Visible Light. *Chem. Mater.* **2008**, *20*, 110–117.

[212] L. Amirav, A. P. Alivisatos. Photocatalytic Hydrogen Production with Tunable Nanorod Heterostructures. *J. Phys. Chem. Lett.* **2010**, *1*, 1051–1054.

[213] M. J. Berr, F. F. Schweinberger, M. Döblinger, K. E. Sanwald, C. Wolff, J. Breimeier, A. S. Crampton, C. J. Ridge, M. Tschurl, U. Heiz, F. Jäckel, J. Feldmann. Size-Selected Subnanometer Cluster Catalysts on Semiconductor Nanocrystal Films for Atomic Scale Insight into Photocatalysis. *Nano Lett.* **2012**, *12*, 5903–5906.

[214] F. F. Schweinberger, M. J. Berr, M. Döblinger, C. Wolff, K. E. Sanwald, A. S. Crampton, C. J. Ridge, F. Jäckel, J. Feldmann, M. Tschurl, U. Heiz. Cluster Size Effects in the Photocatalytic Hydrogen Evolution Reaction. *J. Am. Chem. Soc.* **2013**, *135*, 13262–13265.

[215] S. M. Kim, S. J. Lee, S. H. Kim, S. Kwon, K. J. Yee, H. Song, G. A. Somorjai, J. Y. Park. Hot Carrier-Driven Catalytic Reactions on Pt-CdSe-Pt Nanodumbbells and Pt/GaN under Light Irradiation. *Nano Lett.* **2013**, *13*, 1352–1358.

[216] P. Wang, B. Huang, Y. Dai, M.-H. Whangbo. Plasmonic photocatalysts: harvesting visible light with noble metal nanoparticles. *Phys. Chem. Chem. Phys.* **2012**, *14*, 9813–9825.

[217] Z. Liu, W. Hou, P. Pavaskar, M. Aykol, S. B. Cronin. Plasmon Resonant Enhancement of Photocatalytic Water Splitting Under Visible Illumination. *Nano Letters* **2011**, *11*, 1111–1116.

[218] D. M. Schaadt, B. Feng, E. T. Yu. Enhanced semiconductor optical absorption via surface plasmon excitation in metal nanoparticles. *Appl. Phys. Lett.* **2005**, *86*, 063106 (3 pages).

[219] H. A. Atwater, A. Polman. Plasmonics for improved photovoltaic devices. *Nature Mater.* **2010**, *9*, 205–213.

[220] K. K. Haldar, G. Sinha, J. Lahtinen, A. Patra. Hybrid Colloidal Au-CdSe Pentapod Heterostructures Synthesis and Their Photocatalytic Properties. *ACS Appl. Mater. Interfaces* **2012**, *4*, 6266–6272.

[221] I. Sunagawa, *Crystals: Growth, Morphology and Perfection*, Cambrige University Press, Cambridge, **2005**.

[222] I. V. Markov, *Crystal growth for beginners: fundamentals of nucleation, crystal growth, and epitaxy 2nd ed.*, World Scientific, Hackensack, **2008**.

[223] L. Carbone, S. Kudera, C. Giannini, G. Ciccarella, R. Cingolani, P. D. Cozzoli, L. Manna. Selective reactions on the tips of colloidal semiconductor nanorods. *J. Mater. Chem.* **2006**, *16*, 3952–3956.

[224] A. Perro, S. Reculusa, S. Ravaine, E. Bourgeat-Lami, E. Duguet. Design and synthesis of Janus micro- and nanoparticles. *J. Mater. Chem.* **2005**, *15*, 3745–3760.

[225] C. O'Sullivan, R. D. Gunning, C. A. Barrett, A. Singh, K. M. Ryan. Size controlled gold tip growth onto II-VI nanorods. *J. Mater. Chem.* **2010**, *20*, 7875–7880.

[226] M. G. Alemseghed, T. P. A. Ruberu, J. Vela. Controlled Fabrication of Colloidal Semiconductor-Metal Hybrid Heterostructures: Site Selective Metal Photo Deposition. *Chem. Mater.* **2011**, *23*, 3571–3579.

[227] G. Dukovic, M. G. Merkle, J. H. Nelson, S. M. Hughes, A. P. Alivisatos. Photodeposition of Pt on Colloidal CdS and CdSe/CdS Semiconductor Nanostructures. *Adv. Mater.* **2008**, *20*, 4306–4311.

[228] T. Mokari, C. G. Sztrum, A. Salant, E. Rabani, U. Banin. Formation of asymmetric one-sided metal-tipped semiconductor nanocrystal dots and rods. *Nature Mater.* **2005**, *4*, 855–863.

[229] Y. Khalavka, C. Sönnichsen. Growth of Gold Tips onto Hyperbranched CdTe Nanostructures. *Adv. Mater.* **2008**, *20*, 588–591.

[230] C. O'Sullivan, S. Ahmed, K. M. Ryan. Gold tip formation on perpendicularly aligned semiconductor nanorod assemblies. *J. Mater. Chem.* **2008**, *18*, 5218–5222.

[231] D. R. Lide (Ed.), *CRC Handbook of Chemistry and Physics 89th ed.*, CRC Press, Boca Raton, FL, **2008**, pp. 8-20–8-29.

[232] S. G. Bratsch. Standard Electrode Potentials and Temperature Coefficients in Water at 298.15 K. *J. Phys. Chem. Ref. Data* **1989**, *18*, 1–21.

[233] E. W. Kern, A., *ICDD Grant-in-Aid*, Tech. Rep., Mineralogisch-Petrograph. Inst., Univ. Heidelberg, Germany, **1993**.

[234] H. Swanson, E. Tatge. Standard X-ray diffraction powder patterns. *Natl. Bur. Stand. (U.S), Circ. 539* **1953**, *1*, 12; JCPDS # 00–004–0783.

[235] H. Swanson, E. Tatge. Standard X-ray diffraction powder patterns. *Natl. Bur. Stand. (U.S), Circ. 539* **1953**, *1*, 1–95; JCPDS # 00–004–0784.

[236] J. Häglund, A. Fernández Guillermet, G. Grimvall, M. Körling. Theory of bonding in transition-metal carbides and nitrides. *Phys. Rev. B* **1993**, *48*, 11685–11691; JCPDS # 01–088–2343.

[237] H. Li, M. Zanella, A. Genovese, M. Povia, A. Falqui, C. Giannini, L. Manna. Sequential Cation Exchange in Nanocrystals: Preservation of Crystal Phase and Formation of Metastable Phases. *Nano Lett.* **2011**, *11*, 4964–4970.

[238] M. Casavola, M. A. van Huis, S. Bals, K. Lambert, Z. Hens, D. Vanmaekelbergh. Anisotropic Cation Exchange in PbSe/CdSe Core/Shell Nanocrystals of Different Geometry. *Chem. Mater.* **2012**, *24*, 294–302.

[239] E. M. Chan, M. A. Marcus, S. Fakra, M. ElNaggar, R. A. Mathies, A. P. Alivisatos. Millisecond Kinetics of Nanocrystal Cation Exchange Using Microfluidic X-ray Absorption Spectroscopy. *J. Phys. Chem. A* **2007**, *111*, 12210–12215.

[240] J. B. Rivest, P. K. Jain. Cation exchange on the nanoscale: an emerging technique for new material synthesis, device fabrication, and chemical sensing. *Chem. Soc. Rev.* **2013**, *42*, 89–96.

[241] B. J. Beberwyck, Y. Surendranath, A. P. Alivisatos. Cation Exchange: A Versatile Tool for Nanomaterials Synthesis. *J. Phys. Chem. C* **2013**, *117*, 19759–19770.

[242] L. Dloczik, R. Könenkamp. Nanostructure Transfer in Semiconductors by Ion Exchange. *Nano Lett.* **2003**, *3*, 651–653.

[243] J. Park, H. Zheng, Y.-w. Jun, A. P. Alivisatos. Hetero-Epitaxial Anion Exchange Yields Single-Crystalline Hollow Nanoparticles. *J. Am. Chem. Soc.* **2009**, *131*, 13943–13945.

[244] M. Saruyama, Y.-G. So, K. Kimoto, S. Taguchi, Y. Kanemitsu, T. Teranishi. Spontaneous Formation of Wurzite-CdS/Zinc Blende-CdTe Heterodimers through a Partial Anion Exchange Reaction. *J. Am. Chem. Soc.* **2011**, *133*, 17598–17601.

[245] D. H. Son, S. M. Hughes, Y. Yin, A. Paul Alivisatos. Cation Exchange Reactions in Ionic Nanocrystals. *Science* **2004**, *306*, 1009–1012.

[246] R. G. Parr, R. G. Pearson. Absolute hardness: companion parameter to absolute electronegativity. *J. Am. Chem. Soc.* **1983**, *105*, 7512–7516.

[247] J. M. Luther, H. Zheng, B. Sadtler, A. P. Alivisatos. Synthesis of PbS Nanorods and Other Ionic Nanocrystals of Complex Morphology by Sequential Cation Exchange Reactions. *J. Am. Chem. Soc.* **2009**, *131*, 16851–16857.

[248] S. E. Wark, C.-H. Hsia, D. H. Son. Effects of Ion Solvation and Volume Change of Reaction on the Equilibrium and Morphology in Cation-Exchange Reaction of Nanocrystals. *J. Am. Chem. Soc.* **2008**, *130*, 9550–9555.

[249] R. D. Shannon. Revised effective ionic radii and systematic studies of interatomic distances in halides and chalcogenides. *Acta Crystallogr. Sect. A* **1976**, *32*, 751–767.

[250] F. E. Wagner, P. Palade, J. Friedl, G. Filoti, N. Wang. 197 Au Mössbauer study of gold selenide, AuSe. *J. Phys.: Conf. Ser.* **2010**, *217*, 012039.

[251] K. Schubert, H. Breimer, W. Burkhardt, E. Günzel, R. Haufler, H. Lukas, H. Vetter, J. Wegst, M. Wilkens. Einige strukturelle Ergebnisse an metallischen Phasen II. *Naturwissenschaften* **1957**, *44*, 229–230.

[252] J. Akhtar, R. F. Mehmood, M. A. Malik, N. Iqbal, P. O'Brien, J. Raftery. A novel single source precursor: [bis(N,N-diethyl-N[prime or minute]-naphthoyl-selenoureato)palladium(ii)] for palladium selenide thin films and nanoparticles. *Chem. Commun.* **2011**, *47*, 1899–1901.

[253] S. Geller. The crystal structure of $Pd_{17}Se_{15}$. *Acta Cryst.* **1962**, *15*, 713–721.

[254] F. Grønvold, E. Røst. On the Sulfides, Selenides, and Tellurides of Palladium. *Acta Chem. Scand.* **1956**, *10*, 1620–1634.

[255] F. Grønvold, E. Røst. The crystal structure of $PdSe_2$ and PdS_2. *Acta Cryst.* **1957**, *10*, 329–331.

[256] Wiegers, *Private Communication*, Tech. Rep., University Bloemsingel 10, Groningen, The Netherlands, **1972**.

[257] P. Matkovic, K. Schubert. Kristallstruktur von Pt_5Se_4. *J. Less Common Met.* **1977**, *55*, 185–190.

[258] S. Kudera, L. Carbone, M. F. Casula, R. Cingolani, A. Falqui, E. Snoeck, W. J. Parak, L. Manna. Selective Growth of PbSe on One or Both Tips of Colloidal Semiconductor Nanorods. *Nano Lett.* **2005**, *5*, 445–449.

[259] J. Yang, X. Chen, F. Ye, C. Wang, Y. Zheng, J. Yang. Core-shell CdSe@Pt nanocomposites with superior electrocatalytic activity enhanced by lateral strain effect. *J. Mater. Chem.* **2011**, *21*, 9088–9094.

[260] C. A. Leatherdale, W.-K. Woo, F. V. Mikulec, M. G. Bawendi. On the Absorption Cross Section of CdSe Nanocrystal Quantum Dots. *J. Phys. Chem. B* **2002**, *106*, 7619–7622.

[261] W. Lu, B. Wang, J. Zeng, X. Wang, S. Zhang, J. G. Hou. Synthesis of Core/Shell Nanoparticles of Au/CdSe via Au-Cd Bialloy Precursor. *Langmuir* **2005**, *21*, 3684–3687.

[262] K. Torigoe, K. Esumi. Preparation of colloidal gold by photoreduction of tetracyanoaurate(1-)-cationic surfactant complexes. *Langmuir* **1992**, *8*, 59–63.

[263] K. Yamamoto, S. Inada. Liquid-Liquid Distribution of Ion Associates Tetrahalogenoaurate(III) with Quaternary Ammonium Counter Ions. *Anal. Sci.* **1995**, *11*, 643–649.

[264] R. Puddephatt, *The Chemistry of Gold*, Elsevier, Amsterdam, **1978**.

[265] N. Wiberg, *Lehrbuch der Anoganischen Chemie*, de Gruyter, Berlin, **2007**.

[266] J. Yang, J. Y. Ying. A general phase-transfer protocol for metal ions and its application in nanocrystal synthesis. *Nature Mater.* **2009**, *8*, 683–689.

[267] Z. Huo, C.-k. Tsung, W. Huang, X. Zhang, P. Yang. Sub-Two Nanometer Single Crystal Au Nanowires. *Nano Lett.* **2008**, *8*, 2041–2044.

[268] W. Kim, S. J. Lim, S. Jung, S. K. Shin. Binary Amine-Phosphine Passivation of Surface Traps on CdSe Nanocrystals. *J. Phys. Chem. C* **2010**, *114*, 1539–1546.

[269] R. Quiñones, A. Raman, E. S. Gawalt. An approach to differentiating between multi- and monolayers using MALDI-TOF MS. *Surf. Interface Anal.* **2007**, *39*, 593–600.

[270] *IR-spectrum of 1-dodecanethiol*, Spectral Database for Organic Compounds (SDBS), [Online], SDBS No.: 10643. Last accessed 1 March, **2014**. http://sdbs.db.aist.go.jp/sdbs/cgi-bin/direct_frame_disp.cgi?sdbsno=10643.

[271] E. Khon, N. N. Hewa-Kasakarage, I. Nemitz, K. Acharya, M. Zamkov. Tuning the Morphology of Au/CdS Nanocomposites through Temperature-Controlled Reduction of Gold-Oleate Complexes. *Chem. Mater.* **2010**, *22*, 5929–5936.

[272] A. Halder, N. Ravishankar. Gold Nanostructures from Cube-Shaped Crystalline Intermediates. *J. Phys. Chem. B* **2006**, *110*, 6595–6600.

[273] G. Corthey, L. J. Giovanetti, J. M. Ramallo-López, E. Zelaya, A. A. Rubert, G. A. Benitez, F. G. Requejo, M. H. Fonticelli, R. C. Salvarezza. Synthesis and Characterization of Gold@Gold(I)-Thiomalate Core@Shell Nanoparticles. *ACS Nano* **2010**, *4*, 3413–3421.

[274] P. J. G. Goulet, A. Leonardi, R. B. Lennox. Oxidation of Gold Nanoparticles by Au(III) Complexes in Toluene. *J. Phys. Chem. C* **2012**, *116*, 14096–14102.

[275] P. Williams. Motion of small gold clusters in the electron microscope. *Appl. Phys. Lett.* **1987**, *50*, 1760–1762.

[276] A. V. Naumkin, A. Kraut-Vass, S. Gaarenstroom, C. J. Powell, *NIST X-ray Photoelectron Spectroscopy Database, NIST Standard Reference Database 20, Version 4.1*, [Online], **2012**, Last accessed: 21 March, **2014**. http://srdata.nist.gov/xps/.

[277] M.-C. Bourg, A. Badia, R. B. Lennox. Gold-Sulfur Bonding in 2D and 3D Self-Assembled Monolayers: XPS Characterization. *J. Phys. Chem. B* **2000**, *104*, 6562–6567.

[278] P. Zhang, T. K. Sham. X-Ray Studies of the Structure and Electronic Behavior of Alkanethiolate-Capped Gold Nanoparticles: The Interplay of Size and Surface Effects. *Phys. Rev. Lett.* **2003**, *90*, 245502.

[279] H. Kitagawa, N. Kojima, T. Nakajima. Studies of mixed-valence states in three-dimensional halogen-bridged gold compounds, Cs_2AuAuX_6, (X = Cl, Br or I). Part 2. X-Ray photoelectron spectroscopic study. *J. Chem. Soc., Dalton Trans.* **1991**, 3121–3125.

[280] A. Gole, C. J. Murphy. Seed-Mediated Synthesis of Gold Nanorods: Role of the Size and Nature of the Seed. *Chem. Mater.* **2004**, *16*, 3633–3640.

[281] H. Hakkinen. The gold-sulfur interface at the nanoscale. *Nature Chem.* **2012**, *4*, 443–455.

[282] R. de Paiva, R. Di Felice. Atomic and Electronic Structure at Au/CdSe Interfaces. *ACS Nano* **2008**, *2*, 2225–2236.

[283] Y. Chen, R. E. Palmer, J. P. Wilcoxon. Sintering of Passivated Gold Nanoparticles under the Electron Beam. *Langmuir* **2006**, *22*, 2851–2855.

[284] F. Baletto, C. Mottet, R. Ferrando. Molecular dynamics simulations of surface diffusion and growth on silver and gold clusters. *Surf. Sci.* **2000**, *446*, 31–45.

[285] C. Shen, C. Hui, T. Yang, C. Xiao, J. Tian, L. Bao, S. Chen, H. Ding, H. Gao. Monodisperse Noble-Metal Nanoparticles and Their Surface Enhanced Raman Scattering Properties. *Chem. Mater.* **2008**, *20*, 6939–6944.

[286] R. Keim (Ed.), *Gmelins Handbuch der anorganischen Chemie - Silber Teil 3B 8th ed.*, Verl. Chemie, Weinheim, **1973**, p. 153.

[287] T. Smit, E. Venema, J. Wiersma, G. Wiegers. Phase transitions in silver gold chalcogenides. *J. Solid State Chem.* **1970**, *2*, 309–312.

[288] R. D. Robinson, B. Sadtler, D. O. Demchenko, C. K. Erdonmez, L.-W. Wang, A. P. Alivisatos. Spontaneous Superlattice Formation in Nanorods Through Partial Cation Exchange. *Science* **2007**, *317*, 355–358.

[289] Y. Cai, D. Wolfkühler, A. Myalitsin, J. Perlich, A. Meyer, C. Klinke. Tunable Electrical Transport through Annealed Monolayers of Monodisperse Cobalt-Platinum Nanoparticles. *ACS Nano* **2011**, *5*, 67–72.

[290] R. J. Meyer, E. H. E. Pietsch, *Gmelin handbook of inorganic and organometallic chemistry - Pd Main Vol. B2 8th ed.*, Gmelin Institute for Inorganic Chemistry of the Max-Planck-Society for the Advancement of Science (Eds.), Verl. Chemie, Berlin, **1989**, pp. 241-242.

[291] V. L. Colvin, M. C. Schlamp, A. P. Alivisatos. Light-emitting diodes made from cadmium selenide nanocrystals and a semiconducting polymer. *Nature* **1994**, *370*, 354–357.

[292] K.-S. Cho, E. K. Lee, W.-J. Joo, E. Jang, T.-H. Kim, S. J. Lee, S.-J. Kwon, J. Y. Han, B.-K. Kim, B. L. Choi, J. M. Kim. High-performance crosslinked colloidal quantum-dot light-emitting diodes. *Nature Photon.* **2009**, *3*, 341–345.

[293] R. Debnath, O. Bakr, E. H. Sargent. Solution-processed colloidal quantum dot photovoltaics: A perspective. *Energy Environ. Sci.* **2011**, *4*, 4870–4881.

[294] F. J. Ibañez, F. P. Zamborini. Chemiresistive Sensing with Chemically Modified Metal and Alloy Nanoparticles. *Small* **2012**, *8*, 174–202.

[295] M. Franke, T. Koplin, U. Simon. Metal and Metal Oxide Nanoparticles in Chemiresistors: Does the Nanoscale Matter? *Small* **2006**, *2*, 36–50.

[296] C. B. Murray, C. R. Kagan, M. G. Bawendi. Synthesis and characterization of monodisperse nanocrystals and close-packed nanocrystal assemblies. *Annu. Rev. Mater. Sci.* **2000**, *30*, 545–610.

[297] V. Aleksandrovic, D. Greshnykh, I. Randjelovic, A. Frömsdorf, A. Kornowski, S. V. Roth, C. Klinke, H. Weller. Preparation and Electrical Properties of Cobalt-Platinum Nanoparticle Monolayers Deposited by the Langmuir-Blodgett Technique. *ACS Nano* **2008**, *2*, 1123–1130.

[298] S. Sze, K. K. Ng, *Physics of Semoconductor Devices 3rd ed.*, John Wiley & Sons, Inc., Hoboken, New Jersey, **2007**, chapter 3: Metal-Semiconductor Contacts, pp. 134–196.

[299] A. Zabet-Khosousi, A.-A. Dhirani. Charge Transport in Nanoparticle Assemblies. *Chem. Rev.* **2008**, *108*, 4072–4124.

[300] D. S. Ginger, N. C. Greenham, *Electrical Properties of Semiconductor Nanocrystals* in *Semiconductor and metal nanocrystals Synthesis and electronic and optical properties*, Vol. 87 of *Optical Engineering*, V. Klimov (Ed.), Marcel Dekker Inc., New York, U.S.A., **2003**.

[301] C.-W. Jiang, M. A. Green. Silicon quantum dot superlattices: Modeling of energy bands, densities of states, and mobilities for silicon tandem solar cell applications. *J. Appl. Phys.* **2006**, *99*, 114902.

[302] O. L. Lazarenkova, A. A. Balandin. Electron and phonon energy spectra in a three-dimensional regimented quantum dot superlattice. *Phys. Rev. B* **2002**, *66*, 245319.

[303] M. Drndic, M. V. Jarosz, N. Y. Morgan, M. A. Kastner, M. G. Bawendi. Transport properties of annealed CdSe colloidal nanocrystal solids. *J. Appl. Phys.* **2002**, *92*, 7498–7503.

[304] D. Steiner, D. Azulay, A. Aharoni, A. Salant, U. Banin, O. Millo. Photoconductivity in aligned CdSe nanorod arrays. *Phys. Rev. B* **2009**, *80*, 195308.

[305] Y. Liu, M. Gibbs, J. Puthussery, S. Gaik, R. Ihly, H. W. Hillhouse, M. Law. Dependence of Carrier Mobility on Nanocrystal Size and Ligand Length in PbSe Nanocrystal Solids. *Nano Lett.* **2010**, *10*, 1960–1969.

[306] M. V. Jarosz, V. J. Porter, B. R. Fisher, M. A. Kastner, M. G. Bawendi. Photoconductivity studies of treated CdSe quantum dot films exhibiting increased exciton ionization efficiency. *Phys. Rev. B* **2004**, *70*, 195327.

[307] D. Yu, C. Wang, P. Guyot-Sionnest. n-Type Conducting CdSe Nanocrystal Solids. *Science* **2003**, *300*, 1277–1280.

[308] M. V. Kovalenko, M. Scheele, D. V. Talapin. Colloidal Nanocrystals with Molecular Metal Chalcogenide Surface Ligands. *Science* **2009**, *324*, 1417–1420.

[309] S. V. Voitekhovich, D. V. Talapin, C. Klinke, A. Kornowski, H. Weller. CdS Nanoparticles Capped with 1-Substituted 5-Thiotetrazoles: Synthesis, Characterization, and Thermolysis of the Surfactant. *Chem. Mater.* **2008**, *20*, 4545–4547.

[310] A. W. Wills, M. S. Kang, A. Khare, W. L. Gladfelter, D. J. Norris. Thermally Degradable Ligands for Nanocrystals. *ACS Nano* **2010**, *4*, 4523–4530.

[311] J. Lauth, J. Marbach, A. Meyer, S. Dogan, C. Klinke, A. Kornowski, H. Weller. Virtually Bare Nanocrystal Surfaces: Significantly Enhanced Electrical Transport in $CuInSe_2$ and $CuIn_{1-x}Ga_xSe_2$ Thin Films upon Ligand Exchange with Thermally Degradable 1-Ethyl-5-Thiotetrazole. *Adv. Funct. Mater.* **2013**, *24*, 1081–1088.

[312] D. S. Ginger, N. C. Greenham. Charge injection and transport in films of CdSe nanocrystals. *J. Appl. Phys.* **2000**, *87*, 1361–1368.

[313] P. W. Anderson. Absence of Diffusion in Certain Random Lattices. *Phys. Rev.* **1958**, *109*, 1492–1505.

[314] K. Blech, K. Homberger, U. Simon, *Electrical Properties of Nanoparticles* in *Nanoparticles: From Theory to Application 2nd ed.*, G. Schmid (Ed.), Wiley VCH, Weinheim, **2010**, pp. 401–454.

[315] N. Y. Morgan, C. A. Leatherdale, M. Drndić, M. V. Jarosz, M. A. Kastner, M. Bawendi. Electronic transport in films of colloidal CdSe nanocrystals. *Phys. Rev. B* **2002**, *66*, 075339.

[316] W. J. Parak, L. Manna, F. C. Simmel, D. Gerion, A. P. Alivisatos, *Quantum Dots* in *Nanoparticles: From Theory to Application 2nd ed.*, G. Schmid (Ed.), Wiley-VCH, Weinheim, **2010**.

[317] G. Schön, *Single Electron Tunneling* in *Quantum Transport and Dissipation*, T. Dittrich, G. Hänggi, P. Ingold, B. Kramer, G. Schön, W. Zwerger (Eds.), VCH Verlag, **1997**.

[318] M. A. Kastner. Artificial Atoms. *Phys. Today* **1993**, *46*, 24–31.

[319] K. MacLean, S. Amasha, I. P. Radu, D. M. Zumbühl, M. A. Kastner, M. P. Hanson, A. C. Gossard. Energy-Dependent Tunneling in a Quantum Dot. *Phys. Rev. Lett.* **2007**, *98*, 036802.

[320] D. L. Klein, R. Roth, A. K. L. Lim, A. P. Alivisatos, P. L. McEuen. A single-electron transistor made from a cadmium selenide nanocrystal. *Nature* **1997**, *389*, 699–701.

[321] W. Heywang, *Amorphe Halbleiter* in *Amorphe und polykristalline Halbleiter*, Vol. 18 of *Halbleiter-Elektronik*, Springer Berlin Heidelberg, **1984**, pp. 21–76.

[322] A. Miller, E. Abrahams. Impurity Conduction at Low Concentrations. *Phys. Rev.* **1960**, *120*, 745–755.

[323] N. F. Mott, *Conduction in Non-Crystalline Materials 2nd ed.*, Oxford Clarendon Press, Oxford, Great Britain, **1993**.

[324] M. S. Kang, A. Sahu, D. J. Norris, C. D. Frisbie. Size-Dependent Electrical Transport in CdSe Nanocrystal Thin Films. *Nano Lett.* **2010**, *10*, 3727–3732.

[325] C. A. Neugebauer, M. B. Webb. Electrical Conduction Mechanism in Ultrathin, Evaporated Metal Films. *J. Appl. Phys.* **1962**, *33*, 74–82.

[326] P. Beecher, A. J. Quinn, E. V. Shevchenko, H. Weller, G. Redmond. Charge Transport in Weakly Coupled CoPt$_3$ Nanocrystal Assemblies. *J. Phys. Chem. B* **2004**, *108*, 9564–9567.

[327] D. Greshnykh, A. Froömsdorf, H. Weller, C. Klinke. On the Electric Conductivity of Highly Ordered Monolayers of Monodisperse Metal Nanoparticles. *Nano Lett.* **2009**, *9*, 473–478.

[328] B. Abeles, P. Sheng, M. Coutts, Y. Arie. Structural and electrical properties of granular metal films. *Adv. Phys.* **1975**, *24*, 407–461.

[329] N. F. Mott. Conduction in non-crystalline materials. *Philos. Mag.* **1969**, *19*, 835–852.

[330] A. L. Efros, B. I. Shklovskii. Coulomb gap and low temperature conductivity of disordered systems. *J. Phys. C* **1975**, *8*, L49.

[331] D. Yu, C. Wang, B. L. Wehrenberg, P. Guyot-Sionnest. Variable Range Hopping Conduction in Semiconductor Nanocrystal Solids. *Phys. Rev. Lett.* **2004**, *92*, 216802.

[332] C. A. Leatherdale, C. R. Kagan, N. Y. Morgan, S. A. Empedocles, M. A. Kastner, M. G. Bawendi. Photoconductivity in CdSe quantum dot solids. *Phys. Rev. B* **2000**, *62*, 2669–2680.

[333] A. Persano, G. Leo, L. Manna, A. Cola. Charge carrier transport in thin films of colloidal CdSe quantum rods. *J. Appl. Phys.* **2008**, *104*, 074306.

[334] T. D. Krauss, S. O'Brien, L. E. Brus. Charge and Photoionization Properties of Single Semiconductor Nanocrystals. *J. Phys. Chem. B* **2001**, *105*, 1725–1733.

[335] F. Leonard, A. A. Talin. Electrical contacts to one- and two-dimensional nanomaterials. *Nature Nanotech.* **2011**, *6*, 773–783.

[336] J. Tersoff. Schottky Barrier Heights and the Continuum of Gap States. *Phys. Rev. Lett.* **1984**, *52*, 465–468.

[337] D. Steiner, T. Mokari, U. Banin, O. Millo. Electronic Structure of Metal-Semiconductor Nanojunctions in Gold CdSe Nanodumbbells. *Phys. Rev. Lett.* **2005**, *95*, 056805.

[338] U. Landman, R. N. Barnett, A. G. Scherbakov, P. Avouris. Metal-Semiconductor Nanocontacts: Silicon Nanowires. *Phys. Rev. Lett.* **2000**, *85*, 1958–1961.

[339] D. O. Demchenko, Wang. Localized Electron States near a Metal/Semiconductor Nanocontact. *Nano Lett.* **2007**, *7*, 3219–3222.

[340] H. Woldeghebriel. Electronic Structure of CdSe Nanowires Terminated With Gold Electrodes. *Momona Ethiopian Journal of Science (MEJS)* **2011**, *3*, 5–19.

[341] R. Lavieville, Y. Zhang, E. D. Fabrizio, R. Krahne. Electrical contacts to nanorod networks at different length scales: From macroscale ensembles to single nanorod chains. *Microelectron. Eng.* **2013**, *111*, 185 – 188.

[342] P. Li, A. Lappas, R. Lavieville, Y. Zhang, R. Krahne. CdSe-Au nanorod networks welded by gold domains: a promising structure for nano-optoelectronic components. *J. Nanopart. Res.* **2012**, *14*, 1–5.

[343] J.-S. Lee, M. I. Bodnarchuk, E. V. Shevchenko, D. V. Talapin. "Magnet-in-the-Semiconductor" FePt-PbS and FePt-PbSe Nanostructures: Magnetic Properties, Charge Transport, and Magnetoresistance. *J. Am. Chem. Soc.* **2010**, *132*, 6382–6391.

[344] J. S. Son, J.-S. Lee, E. V. Shevchenko, D. V. Talapin. Magnet-in-the-Semiconductor Nanomaterials: High Electron Mobility in All-Inorganic Arrays of FePt/CdSe and FePt/CdS Core-Shell Heterostructures. *J. Phys. Chem. Lett.* **2013**, *4*, 1918–1923.

[345] R. Chakraborty, F. Greullet, C. George, D. Baranov, E. Di Fabrizio, R. Krahne. Broad spectral photocurrent enhancement in Au-decorated CdSe nanowires. *Nanoscale* **2013**, *5*, 5334–5340.

[346] C. Y. Lau, H. Duan, F. Wang, C. B. He, H. Y. Low, J. K. W. Yang. Enhanced Ordering in Gold Nanoparticles Self-Assembly through Excess Free Ligands. *Langmuir* **2011**, *27*, 3355–3360.

[347] M. Drndic, R. Markov, M. V. Jarosz, M. G. Bawendi, M. A. Kastner, N. Markovic, M. Tinkham. Imaging the charge transport in arrays of CdSe nanocrystals. *Appl. Phys. Lett.* **2003**, *83*, 4008–4010.

[348] D. Greshnykh, *Untersuchungen zum Ladungstransport in Monolagen von Cobalt-Platin-Nanopartikeln*, Ph.D. thesis, Fachbereich Chemie, Universität Hamburg, **2008**.

[349] E. Shaviv, A. Salant, U. Banin. Size Dependence of Molar Absorption Coefficients of CdSe Semiconductor Quantum Rods. *ChemPhysChem* **2009**, *10*, 1028–1031.

[350] J. F. Moulder, W. F. Stickle, P. E. Sobol, K. D. Bomben, *Handbook of X-ray electron spectroscopy*, J. Chastain (Ed.), Perkin-Elmer Corporation, Eden Prairie, **1992**.

[351] I. GESTIS-Stoffdatenbank, [Online], last accessed: 3 April, **2014**. http://www.dguv.de/ifa/Gefahrstoffdatenbanken/GESTIS-Stoffdatenbank/index.jsp.

[352] *1-Chlorooctadecan*, MSDS No. 238368, [Online], Sigma-Aldrich, Steinheim, Germany, 17 May, **2013**, last accessed: 3 April **2014**. http://www.sigmaaldrich.com/MSDS/MSDS/DisplayMSDSPage.do?country=DE&language=de&productNumber=238368&brand=ALDRICH&PageToGoToURL=http%3A%2F%2Fwww.sigmaaldrich.com%2Fcatalog%2Fsearch%3Finterface%3DAll%26term%3D%2520chlorooctadecane%26N%3D0%26focus%3Dproduct%26lang%3Dde%26region%3DDE.

[353] *1,2-Dibromethane*, MSDS No. 112790000, [Online], Acros Organics BVBA, Geel, Belgium, 1 May, **2012**, accessed: 3 April **2014**. http://www.acros.com/Ecommerce/msds.aspx?PrdNr=11279&Country=DE&Language=de.

[354] *1,2-Dichlorobutane*, MSDS No. 105465, [Online], Sigma-Aldrich, Steinheim, Germany, 22 November, **2012**, last accessed: 3 April **2014**. http://www.sigmaaldrich.com/MSDS/MSDS/DisplayMSDSPage.do?country=DE&language=de&productNumber=105465&brand=ALDRICH&PageToGoToURL=http%3A%2F%2Fwww.sigmaaldrich.com%2Fcatalog%2Fsearch%3Finterface%3DAll%26term%3D%2520dichlorobutane%26N%3D0%26focus%3Dproduct%26lang%3Dde%26region%3DDE.

[355] *1,2-Dichloroethane*, MSDS No. 113713, [Online], Merck KGaA, Darmstadt, Germany, 26 March, **2014**, last accessed: 3 April **2014**. http://www.merckmillipore.com/germany/chemicals/1-2-dichlorethan/MDA_CHEM-113713/p_HJCb.s1Ld8gAAAEWp.AfVhTl;sid=i5jro7xdFJqvo-9usCeG9BSdDBv5U9b6s7b19Q_wqexVwDLTp3iICbpjnMNmHpckC-v7v7-EdqDCpDWUKRxeSBMXqmzy53CNuAJTggREM71\-Y5QR8M4H8-DHg?attachments=MSDS.

[356] *1,2-Diiodoethane*, MSDS No. D122807, [Online], Sigma-Aldrich, Steinheim, Germany, 30 April, **2013**, last accessed: 3 April **2014**. http://www.sigmaaldrich.com/MSDS/MSDS/DisplayMSDSPage.do?country=DE&language=de&productNumber=D122807&brand=ALDRICH&PageToGoToURL=http%3A%2F%2Fwww.sigmaaldrich.com%2Fcatalog%2Fproduct%2Faldrich%2Fd122807%3Flang%3Dde.

[357] *n-Dodecyltrimethylammonium bromide*, MSDS No. A10761, [Online], Alfa Aesar, Karlsruhe, Germany, 8 February, **2007**, last accessed: 3 April **2014**. http://www.alfa-ebsoft.com/msds/pdf/german/A10761.PDF.

[358] *n-Dodecyltrimethylammonium chloride*, MSDS No. 409320000, [Online], Acros Or-

Bibliography

ganics BVBA, Geel, Belgium, 3 May, **2012**, last accessed: 3 April **2014**. http://www.acros.com/Ecommerce/msds.aspx?PrdNr=40932&Country=DE&Language=de.

[359] *n-Octadecylphosphonic acid*, MSDS No. 20645, [Online], Alfa Aesar, Karlsruhe, Germany, 24 January, **2007**, last accessed: 3 April **2014**. http://www.alfa.com/content/msds/German/20645.pdf.

[360] *Palladium(II) acetate*, MSDS No. 520764, [Online], Sigma-Aldrich, Steinheim, Germany, 8 October, **2012**, last accessed: 3 April **2014**. http://www.sigmaaldrich.com/MSDS/MSDS/DisplayMSDSPage.do?country=DE&language=de&productNumber=520764&brand=ALDRICH&PageToGoToURL=http%3A%2F%2Fwww.sigmaaldrich.com%2Fcatalog%2Fproduct%2Faldrich%2F520764%3Flang%3Dde.

[361] *Platinum(II) acetylacetonate*, MSDS No. 523038, [Online], Sigma-Aldrich, Steinheim, Germany, 21 September, **2012**, last accessed: 3 April **2014**. http://www.acros.com/Ecommerce/msds.aspx?PrdNr=19758&Country=DE&Language=de.

[362] *Tri-n-octylphosphane*, MSDS No. 117854, [Online], Sigma-Aldrich, Steinheim, Germany, 20 November, **2013**, last accessed: 3 April **2014**. http://www.sigmaaldrich.com/MSDS/MSDS/DisplayMSDSPage.do?country=DE&language=de&productNumber=117854&brand=ALDRICH&PageToGoToURL=http%3A%2F%2Fwww.sigmaaldrich.com%2Fcatalog%2Fproduct%2Faldrich%2F117854%3Flang%3Dde.

[363] *Tri-n-octylphosphane oxide*, MSDS No. 814868, [Online], Merck KGaA, Darmstadt, Germany, 17 May, **2013**, last accessed: 3 April **2014**. http://www.merckmillipore.com/chemicals/de_DE/Merck-DE-Site/EUR/ViewProductDocuments-File?ProductSKU=MDA_CHEM-814868&DocumentType=MSD&DocumentId=&DocumentSource=&Country=GLOBAL&Channel=Merck-DE-Site&Language=EN.

Acknowledgements

I would like to thank PD Dr. Christian Klinke for the opportunity to work in his group and for providing excellent experimental facilities as well as for his support, his commitment to science and the fact that his door was always open. Furthermore, I would like to thank him for the POV-Ray images, the DFT calculations and some SEM micrographs in this work.

I would also like to thank Prof. Dr. Horst Weller for being the second referee.

Special thanks are expressed towards Dr. Neus Gómez Bastús from the Catalan Institute of Nanoscience and Nanotechnology and Dr. Beatriz Hernández Juárez from the Universidad Autonóma and IMDEA Nanoscience Madrid for collaborations including joint publications, a stay in the lab in Madrid, stimulating discussions and motivation during the past years.

I would like to give credit to Dr. Hauke Lehmann and Stefan Lohmeier for electrical transport measurements. Dr. Lehmann also helped by preparing electrode structures, if necessary on weekends and in the evening, and by obtaining scanning electron micrographs.

I would like to thank Andreas Kornowski, Daniela Weinert and Stefan Werner for great HR-TEM micrographs, EDX-measurements and discussions about the possibilities of the analytical methods. Additionally, I am grateful to Andreas Kornowski for *in situ* annealing studies and valuable discussions.

Dr. Fabiola Iacono, Dr. Roberto Otero and Dr. José M. Gallego of the Universidad Autónoma de Madrid and IMDEA Nanoscience Madrid are acknowledged for XPS measurements and deconvolution of the data.

ACKNOWLEDGEMENTS

The assistance of my bachelor student Jan Geweke and the students Mehrdad Shiri and Jonny Proppe, all of whom worked so diligently in the lab, was much appreciated.

Almut Barck is thanked for XRD measurements and the nice introduction to the machine.

I am grateful to Mauricio D. Coderch and Dr. Ursula E. A. Fittschen for TXRF and ICP-OES analyses.

Further thanks go to Leonor de la Cueva from the Universidad Autónoma de Madrid and Dr. Cristina Palencia for collaborations and discussions.

A big thank you is expressed towards the current and former members of the Klinke and Weller groups for the nice and cooperative working atmosphere. Thanks to Dr. Hauke Lehmann, not only more physics and coffee entered our office but also a nice colleague. Apart from many scientific and private discussions and an enjoyable time in the lab, the cultural and culinary experiences shared with Dr. Neus Gómez Bastús, Dr. Yuxue Cai, Dr. Christian Schmidtke, Dr. Van-Huong Tran and Dr. Annette Wurl will be kept in good memory. Equally well remembered will be Mirjam Volkmann for being such a helpful and easy to work with colleague with a good taste in music.

Many thanks go to Verena Schmitt, Dr. Hauke Lehmann and Mirjam Volkmann for their help with LaTeX.

Silke Starodubetz is acknowledged for linguistic suggestions.

I would like to thank Verena Schmitt and Julia Gerick for being such good friends and for sharing our experiences during the PhD studies.

I am more than grateful to my parents and Jan Stöver for their constant support and encouragement.

Curriculum Vitae

- **Personal data**:
  Name: Michaela Meyns
  Year of birth: 1985
  Place of birth: Reinbek

- **University**:

04/2010 - 05/2014	PhD studies in the group of PD Dr. C. Klinke at the Institute of Physical Chemistry, University of Hamburg, Germany
	Topic: Metal-semiconductor hybrid nanoparticles: Halogen induced shape control, hybrid synthesis and electrical transport
04/2010 - 03/2014	Research assistant in the group of PD. Dr. C. Klinke
10/2012	Research stay in the group of Dr. B. Hernández Juárez, IMDEA Nanoscience/ Universidad Autónoma de Madrid, Spain
09/2009 - 03/2010	Diploma thesis in the group of PD Dr. C. Klinke
	Topic: Synthesis of hybrid cadmium selenide-gold nanostructures
10/2004 - 03/2010	Diploma course in Chemistry, University of Hamburg
10/2008 - 12/2008	Research stay in the group of Prof. Dr. J. J. DeVoss, University of Queensland, Brisbane, Australia
	Topic: Synthesis of enantiomerically pure hydroxy fatty acids as substrates for Cytochrome $P450_{BioI}$
11/2006	Pre-diploma in Chemistry

CURRICULUM VITAE

- **School**:
08/1995 - 06/2004	Secondary school education terminated with the *Abitur*, Sachsenwaldschule Gymnasium Reinbek
03/2002	School exchange to Amalfi, Italy
06 - 08/2001	School exchange to Korowa Anglican Girls' School, Melbourne, Australia

- **Work experience**:
04/2010 - 02/2013	Laboratory instructor at the Department of Chemistry, *Advanced Physical Chemistry*
09/2009 - 03/2010	Laboratory instructor at the Department of Chemistry, *Chemistry for medical students*
04/2007 - 07-2009	Tutor for Organic Chemistry for students with chemistry as a minor course

Publications and conference contributions

List of publications

- C. Palencia, K. Lauwaet, L. de la Cueva, M. Acebrón, **M. Meyns**, C. Klinke, J. M. Gallego, R. Otero, B. H. Juárez. Cl-capped CdSe nanocrystals via in-situ generation of chloride anions. *Nanoscale* **2014**, *6*, 6812-6818.

- **M. Meyns**, F. Iacono, C. Palencia, J. Geweke, M. D. Coderch, U. E. A. Fittschen, J. M. Gallego, R. Otero, B. H. Juárez, C. Klinke. Shape Evolution of CdSe Nanoparticles controlled by Halogen Compounds. *Chem. Mater.* **2014**, *26*, 1813-1821.

- F. Iacono, C. Palencia, L. de la Cueva, **M. Meyns**, L. Terracciano, A. Vollmer, M. J. de la Mata, C. Klinke, J. M. Gallego, B. H. Juarez, R. Otero. Interfacing Quantum Dots and Graphitic Surfaces with Chlorine Atomic Ligand. *ACS Nano* **2013**, 2559-2565.

- A. A. Singh, S. N. A. Zulkifli, **M. Meyns**, P. Y. Hayes, J. J. De Voss. Synthesis of highly enantioenriched hydroxy- and dihydroxy-fatty esters: substrate precursors for cytochrome P450$_{BioI}$. *Tetrahedron: Asymmetry* **2011**, *22*, 1709-1719.

- **M. Meyns**, N. G. Bastús, Y. Cai, A. Kornowski, B. H. Juárez, H. Weller, C. Klinke. Growth and reductive transformation of a gold shell around pyramidal cadmium selenide nanocrystals. *J. Mater. Chem.* **2010**, *20*, 10602-10605.

- B. H. Juárez, **M. Meyns**, A. Chanaewa, Y. Cai, C. Klinke, H. Weller. Carbon supported CdSe nanocrystals. *J. Am. Chem. Soc.* **2008**, *130*, 15282-15284.

A bachelor's thesis was conducted and published within the framework of this thesis: J. Geweke, *Synthese pyramidaler CdSe-Nanopartikel und der Einfluss chlorhaltiger Additive auf ihre Form*, Bachelorarbeit, Universität Hamburg, **2011**.

PUBLICATIONS AND CONFERENCE CONTRIBUTIONS

Conference contributions

Contributed talks

- M. Meyns, N. G. Bastús, F. Iacono, L. de le Cueva, C. Alonso, R. Otero, B. H. Juárez, C. Klinke. Synthesis and chemical transformations of semiconductor-metal hybrid nanostructures with applicability in energy conversion and (photo)catalysis. *E-MRS Spring Meeting*, Strasbourg, France, **2013**.

- M. Meyns, N. G. Bastús, Y. Cai, A. Kornowski, B. H. Juárez, H. Weller, C. Klinke. Inorganic Hybrid Nanocrystals: Selective reduction of Au precursors on dipyramidal CdSe nanocrystals, *China Nano 2011*, Beijing, China, **2011**.

- M. Meyns, N. G. Bastús, Y. Cai, A. Kornowski, B. H. Juárez, C. Klinke. Nanoparticle heterostructures: Selective growth of metal dots onto dihexagonal pyramidal CdSe nanocrystals, *Nano 2010*, Rome, Italy, **2010**.

Posters

- M. Meyns, N. G. Bastús, C. Klinke. Morphology control, stability and reactivity of semiconductor-metal hybrid nanostructures. *NaNaX5*, Fuengirola, Spain, **2012**.

- M. Meyns, N. G. Bastús, Y. Cai, A. Kornowski, B. H. Juárez, H. Weller, C. Klinke. Inorganic Hybrid Nanocrystals: Selective growth of Au domains onto pyramidal CdSe nanocrystals. *110th Annual meeting of the German Bunsen Society for Physical Chemistry*, Berlin, Germany, **2011**.

Erklärungen

Hiermit erkläre ich an Eides statt, dass ich diese Arbeit selbstständig und unter ausschließlicher Verwendung der angegebenen Hilfsmittel und Quellen verfasst habe.

Weiterhin erkläre ich, dass von mir keine weiteren Promotionsversuche unternommen worden sind und diese Arbeit nicht in gleicher oder ähnlicher Form einer anderen Prüfungsbehörde vorgelegt wurde.

Michaela Meyns
Hamburg, 7.4.2014.

Erklärungen